KT-558-689

THE ALEXANDER PRINCIPLE

THE ALEXANDER PRINCIPLE

by

WILFRED BARLOW

Illustrated by Gwyneth Cole

LONDON
VICTOR GOLLANCZ LTD
1973

ISBN 0 575 00563 7

First published May 1973
Second impression June 1973

The quotation from *East Coker* by T. S. Eliot, reprinted on page 67, is from *Collected Poems, 1909–1962*, and is used by kind permission of the publishers Faber & Faber.

NOTE

All case histories cited in this book have been so altered as to render their subjects unrecognisable.

Printed in Great Britain by
The Camelot Press Ltd, London and Southampton

CONTENTS

NOTE

Text references to photographs are given with the term "plate" followed by a number and a facing or following page number. Text references to drawings and diagrams are given with the term "figure" or "fig.".

LIST OF PLATES

TO MARJORY

Chapter 1

THE ALEXANDER PRINCIPLE

JUST BEFORE THE war, in 1938, I came across a very remarkable man. Seventeen years later, when I wrote his obituary in *The Times*, I called him a genius, and I still see no reason to go back on this. Unfortunately, Alexander was also somewhat of a rogue. His roguery was innocent and patent for all to see and enjoy—he was an entertainer, trained on the stage, as well as a scientist—but it held up full acceptance of his principle in his lifetime.

There is no need here to enlarge on the complex personality of this delightful man, except to marvel that such an earthy character could have come up with such a principle. The confidence-man who discovers that he has a method for making genuine gold-bricks is up against it if he wishes to convince the scientific world that the method is sound; particularly so if he has no scientific language but only the ways of the ordinary world.

Alexander's gold brick was, in fact, HEALTH. His attempts to convince the scientific world did not succeed during his lifetime, but in recent years his principles have been accepted in almost every sphere of human activity. Men and women who are pre-eminent in their various spheres of medicine, science, education, religion, music, politics, art, architecture, industry, literature, philosophy, psychology, the stage and television have gladly accepted the help that the Alexander principle has to offer them: perhaps more important, younger people who came upon it earlier in life are now showing in their successful life-styles just what it has to offer.

This book is a commentary on the thirty years' work which I have done since I met Alexander. It is motivated by a feeling of urgency that, in the present mess and muddle in which we find ourselves, his principle can provide a new touchstone for most of us. It can not only provide a criterion for evaluating our own actions but a criterion for assessing the actions of other people and the value of new popular trends and new popular leaders. It also provides a way of looking at those well-established and time-honoured dogmas which, in the words of Paracelsus, "are not worth a goose's turd".

Alexander's main problem was to persuade people that he had something to offer. The problem still exists, to some extent, but, since the pendulum is now swinging the other way, the problem for most people will be to get hold of someone who knows about the principle and who has not been brought up in the gold-brick atmosphere which dogged Alexander all his life. He has sometimes been poorly served by his followers and many of them so far have found the brick too hot to handle. It is fair to say that only in recent years has the principle found its rightful place in medicine and education. Sufficient numbers of teachers who have been well trained in the principle are only now beginning to appear.

Just one statistic. Since Alexander started working seventy-five years ago when he was 30, more than one hundred teachers of his method have been trained in this country. Of these, only four have died, including Alexander himself aged 87, and his first assistant, Ethel Webb aged 94: no coronaries, no cancers, no strokes, no rheumatoid arthritis, no discs, no ulcers, no neurological disorders, no severe mental disorder, just occasionally some rather unlikely behaviour; accidents inevitably, but recovery to good functioning and no accident-proneness. By and large, a standard of day-to-day health and happiness which most people encounter only in their earliest years.

This statistic is almost unbelievable. The gold-brick is too good to be true, but it is in fact true, and it is not insufferable arrogance to say that this principle is a MUST. It is by now a plain brute fact that the Alexander principle works. It is also a plain brute fact that over 99% of the population need it but know nothing of it: I hope that, provided the writing of this book does not make me the fifth mortality, I can get across some of its importance to the general reader.

The Principle

So much for preliminary encouragement. Plenty of this is needed if we are to look at things in a completely different way. What does the Alexander Principle say? The Alexander Principle says that

USE AFFECTS FUNCTIONING

The old negro preacher who was asked for the secret of his success, said, "First, ah tells them what ah's gonna say: then ah says it: then ah tells them what ah's just said." So here it is again

USE AFFECTS FUNCTIONING

What sort of use? What sort of functioning? This is what we need to consider.

Use

You are sitting somewhere, perhaps lying somewhere, reading this book. Are you aware of how your hand is holding the book? If you direct your attention to your hand, you will become aware of the pressure of your fingers taking the weight of the book. How are you sitting? Are your knees crossed? Is the weight of your body more on one buttock than the other? Where are your elbows? As you run your eyes over the page, does your head move to alter your eye position, or do just your eyes move? Where are your shoulder-blades? How much muscle tension are you creating in your chest and forearms and generally throughout your body?

USE means the way we use our bodies as we live from moment to moment. Not only when we are moving but when we are keeping still. Not only when we are speaking but when we are thinking. Not only when we are making love but when we are feeling or refusing to feel pleasure. Not only when we are communicating by actual gestures and attitudes but when, unknown to ourselves, our whole bodily mood and disposition tells people what we are like and keeps us that way whether we like it or not. Not only when we are searching and manipulating our surroundings but when we are letting our surroundings manipulate us like puppets: liking what we get, instead of getting what we like.

USE is the theme of this book. Its variety will be described in the chapters which follow.

Functioning

FUNCTIONING is also the theme of this book. All of us are functioning—adequately, inadequately, happily, unhappily, healthily, unhealthily. A few case-histories will give an idea of the sort of wrong FUNCTIONING which can go with wrong USE. It may not be immediately clear why these people's USE was the first thing which needed tackling: it is the purpose of this book to indicate the connection and the remedy.

The Physician

Dr James P, a chest physician, has been worried for some time by increasing depression and a constant pain in his neck. He assuages it with liberal doses of alcohol and by the thanks of his grateful patients. He is a scholarly man who knows all about depression and psychosomatic pains in the neck. His own neck still hurts, and it is getting him down.

About twenty years ago, when he was a timid medical student, he opted for a rather pompous manner which involved straightening his neck, pulling his chin down on to his throat and occasionally belching—the sort of gentle belching which is a common form of parlance in aristocratic clubs, preceded by a slight swallowing of air to provide the necessary ammunition.

A few years later, he refined the head posture to include a deprecatory twist of his head to one side and a puffing-out of his chest in front. A few years later, he was making these movements even when he was alone and sitting quite still: the belching had become a habit, and, in between belching, he tightened his throat and restricted his breathing.

Dr P had already consulted his psychiatric colleagues and he had reluctantly cut down on much of his work since he found it impossible to concentrate. There was not the remotest possibility of him getting rid of his neck pain until his strange muscular usages had been sorted out. Some of his problems are dealt with in Chapter 6 (Use and Disease).

The Student

Jane B, a pretty 19-year-old, had been reading English Literature at one of the new universities. At school she very nearly didn't take her A levels—she was basically a "good" girl, and was one of the two girls in her sixth form who had never smoked nor lost her virginity. After two days at university, she turned up at home saying she couldn't stand it, but reluctantly agreed to go back—her parents later on rather

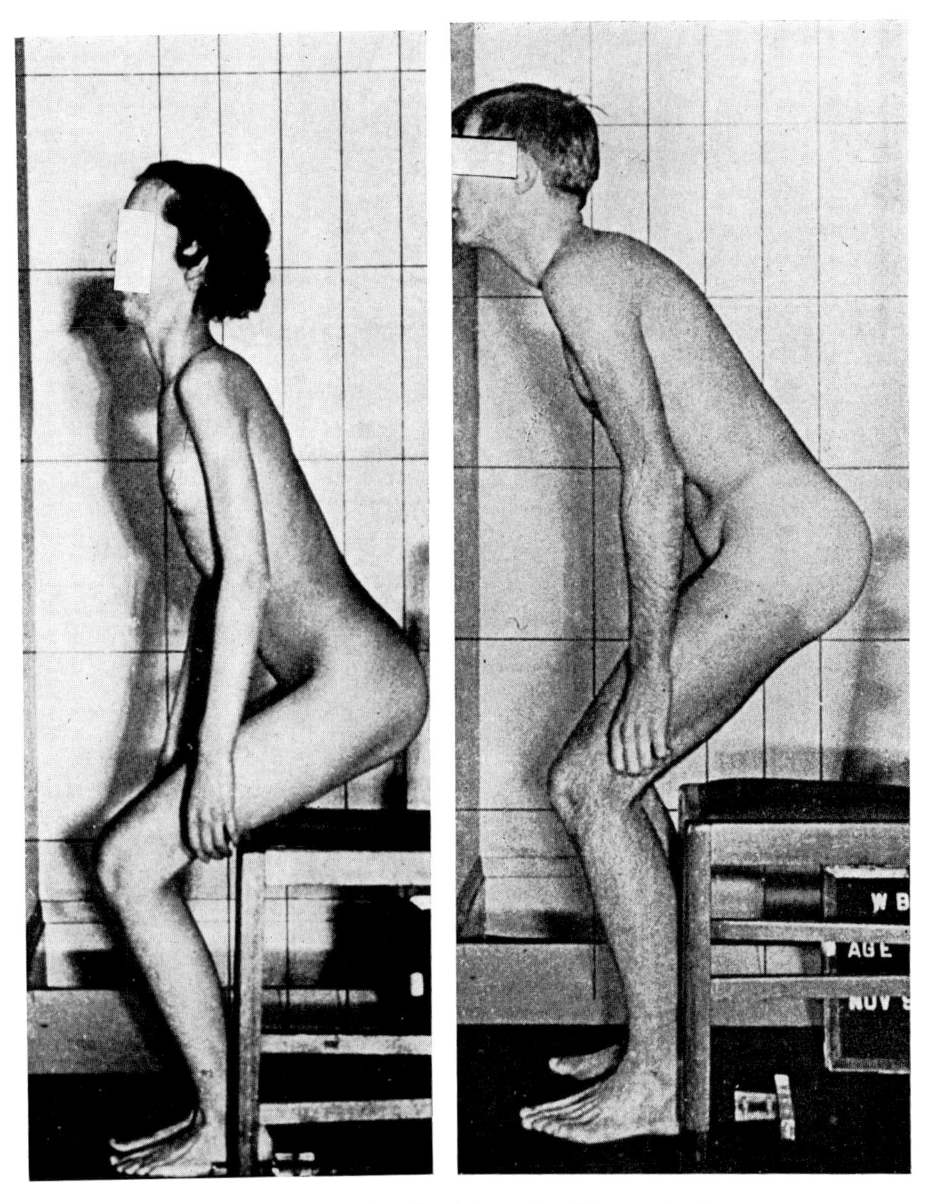

1a 1b Head pulling back into shoulders as in fig. 1.

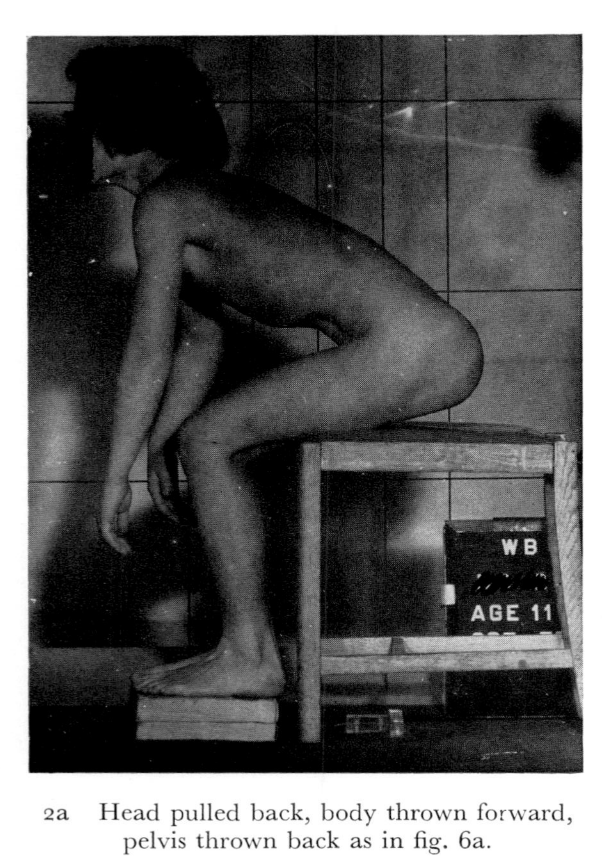

2a Head pulled back, body thrown forward, pelvis thrown back as in fig. 6a.

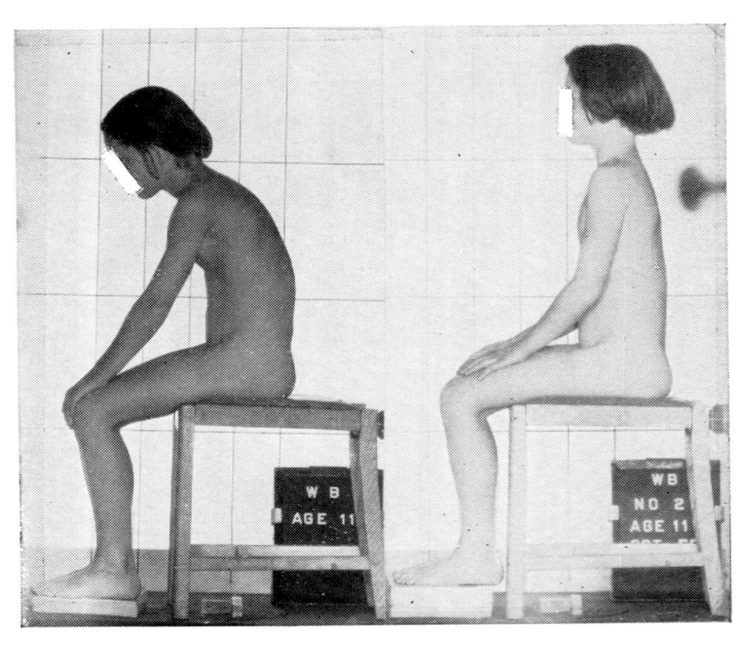

2b and 2c Slump versus balanced USE.

wished that they had abided by their daughter's instinctive rejection.

She found herself painfully unable to talk with other people and she withdrew more and more into herself. At this point, like many others, she could have chosen to fall in with the general permissive scene, and like many others, if her temperament had been suited to it, she might have sickened of it and made stable enjoyable relationships which used the social anti-depressant drugs as an occasional pleasure rather than a constant haven.

Jane B instead cracked up. She wept almost continuously, not just with tears from her eyes, but with an agonised contraction through her whole body. Her stomach contracted, her hands twisted and tensed, her eyes and head dropped down on her chest, and her shoulders lifted up towards her ears. The psychiatrists said it was "reactive depression" and treated her with shock therapy and anti-depressant drugs. It was not until her USE was considered that the breakthrough to improved functioning became possible. The relationship of USE to Mental Functioning is described in Chapters 7 (Mental Health) and 12 (Applying the Principle).

The Journalist

Mrs Hermione X used to be well known in the world of women's magazines—wittily informed and human, identifiably plagued with the problems of her wittily informed and human readers. A good degree had neither kept her in the Shades of Academe nor precipitated her into the world of giggling revolutionaries. At the age of 45, married, with grown-up children and time to spare, she suddenly started to wonder what it was all about. She came to me ostensibly for help with her hobby of flute-playing—her breathing and her fingering were totally unpredictable—but a whole range of psychosomatic symptoms were soon being presented to me for appraisal. It soon became apparent that her main trouble was her sex-life. It has been said that at any given time you are in trouble because you are worrying about your sex life: or you are in real trouble because you are not worrying about your sex-life.

Hermione's problem was fairly clear. She felt constantly very sexy, and could be triggered off into sexual excitement by quite small things: but when she began to engage in actual love-making, she would go cold and dead. The harder she worked towards orgasm, the more irritated and less responsive she became.

The whole range of sexual problems in which muscular usage plays a prime part are dealt with in Chapter 8 (The Psycho-mechanics of Sex) but it should not be thought that these are simply the problems of the tense pelvis—Reich's "Frozen Pelvis". Each of us will have elaborated our own minutely variegated system of muscular usages not only in our pelvis but throughout the whole of our bodies—usages which are suddenly thrown up to demand their share of the picture at the most inconvenient moment, and to interfere with the balanced functioning which we expect from our bodies.

The Schoolboy

Edward P is 11 years old. Two years ago he felt a curious "thumping" at the back of his head, and when his mother became worried by what he told her, he started to get very upset himself. The school doctor could not explain it, nor could a whole battery of neurologists, orthopaedic surgeons and ear, nose and throat specialists who were consulted. The young boy by now was becoming extremely hypochondriacal about it and would argue at length with his mother about how he had said the symptoms felt at such and such a time. Their GP became heartily sick of it all, and a new GP had little extra to offer. A close friend had heard of an osteopath who could do wonders, so the poor little boy went to the osteopath to have his neck cricked week after week, and the "thumping" was quite a convenient reason not to undertake school activities when there was stress around. As so often happens, however—and as may have been the case all along—Edward's pattern of muscular use began to deteriorate in an alarming way.

When he came to see me with his young mother, he could not keep his neck and shoulders still for more than a few seconds, and I was fascinated to see that his mother participated fully in this pattern of muscular twisting and wriggling. They communicated with each other by fractional shifts of muscular adjustment, in which one of them would counter the hints or suggestions of the other by a movement which in its turn had to be countered. This game of muscular ping-pong between the two was quite unconscious and it reminded me of a Jungian psychotherapist who had a habit of establishing rapport with her patients by a series of knowing wriggles and nods of her head which seemed to herself like an exhibition of friendliness, but which must have felt like an irritating intrusion to the patient. The psychotherapist—whom I was treating for a muscle-tension state—said that when I taught her a balanced state of rest, she felt as if she wasn't establishing proper rapport with her patients, although in fact they seemed to find it easier to talk to her.

Edward P was not untypical of boys of his age and such problems of growth are described in Chapter 9 (Personal Growth). By the age of 11, 70% of all boys and girls will already show quite marked muscular and posture deficiencies. Mostly these defects show themselves as passing inefficiencies and difficulties in learning: they become accentuated in emotional situations, and they presage an uneasy adolescence in which childhood faults become blown up into fully fledged defects. By the age of 18, only 5% of the population are free from defects, 15% will have slight defects, 65% will have quite severe defects and 15% will have very severe defects. These figures are based on my published surveys of boys and girls from secondary schools and students from physical training, music and drama colleges, some of whom might reasonably be expected to have a higher physical standard than the rest of the population. It is almost certain that you, the reader of this book, have quite marked defects of which you yourself are unconscious, and which your doctors or teachers or parents did not notice or did not worry about or accepted as an inevitable part of the way you are made.

* * *

These case-histories all say the same thing, that USE affects FUNCTIONING. The physician with his neck-pain, the student with her depression, the journalist with her muscular frigidity, the schoolboy with his habit-spasm—all had been pathetically mishandled by their doctors. Diagnosis in their case had been inadequate, not only because of a mistaken idea of what diagnosis should involve, but because of a failure to observe and to understand what is meant by USE. Chapter 5 (Medical

Diagnosis) and 6 (Use and Disease) show where the Alexander Principle fits into the medical picture, not just in the isolated examples I have given above, but in the whole wide spectrum of ill health as it affects us all.

By the time we reach adult life, if not before, most of us will have developed tension habits which are harmful. The habits at first may show themselves only as trifling inconsistencies of behaviour, or perhaps as occasional muscular pain or clumsiness. Frequently, however, they show themselves as infuriating blockages which prevent us from giving our best just when we most need to, whether it be in the everyday business of personal relations, or in the more exacting situations of competitive sport, public speaking, making music or making love. In any situation, in fact, in which our FUNCTIONING is affected by our USE.

Matthias Alexander

The Alexander Principle sounds at first deceptively simple. I have called it the Alexander Principle because, as far as I know, he was the first person to state it, and I have endeavoured in this book to give an account which will be helpful to someone who has never heard of him.

Much that is inaccurate has been written about Matthias Alexander. I knew him intimately for over ten years, married into his family and have edited the *Alexander Journal* for many years. In his later years, he asked my wife and me to be responsible for the future of his work, and at his request, I founded the Alexander Society of Teachers, with this in view. I know as well as anyone his personal idiosyncrasies—they wear extremely well now that he is dead, however much they may have upset people in his lifetime. I suspect that if Alexander were alive today he would be found to be speaking very clearly to our present condition: his predictions of our present personal and social unhappinesses have come about very much as he foretold.

I should perhaps briefly mention the contact which I myself had with him and his ideas before and after the war. He had come to London in 1904, aged 34, from Sydney, where he had been the director of the "Sydney Dramatic and Operatic Conservatorium". His concept of USE was not very clearly formulated at that time, but between 1904 and 1955 he published four books, of which the shortest and perhaps most to the point was *The Use of the Self* (Methuen, 1932). This book aroused considerable interest on the part of doctors and teachers and many others, especially in the 1930's. For example, George Bernard Shaw, Aldous Huxley, Stafford Cripps, and Archbishop William Temple were his pupils.

I first heard of him through reading Aldous Huxley's *Ends and Means* in 1937. Almost everything he was saying made sense to me, and I decided to study under him to learn to teach his methods. We struck a very close bond and he certificated me as a teacher in 1940, just after the outbreak of war.

Working in war-time London became difficult and he was evacuated to America with his school in the summer of 1940. It looked at that time as though his work and his principle could easily become lost—nearly all of his teachers were in the services. I myself spent a few boring years as a Regimental medical officer seeing little either of the enemy or medicine. It did, however, give me the opportunity to carry out research on large groups of young men and women who were under great emotional and physical stress: and in the process to confirm many of Alexander's observations on USE.

The widespread application and importance of his principle was not immediately obvious in the first half of this century. He had found it at first in his study of the *act of speaking*, and he had made the fairly trite observation that the way people use their muscles affects the way their voices function. Trite, because many schools of speech training, speech therapy and drama now devote themselves to just such a study of the mechanics of voice function. What was not trite about his observation was the minuteness of his analysis of the physical and psychological factors which are involved in USE, and his realisation that by his method of detailed analysis, a large number of psychological and physical disorders—disorders unconnected with the voice—would appear in a completely new light. He realised that in his lifetime he had only touched the fringe of the new possibilities which his approach had opened up.

An Evolutionary Hypothesis

The Alexander Principle is a hypothesis: it is not claimed as an established, absolute truth, but as a new way of looking at things, a new way of organising oneself. It may be that it will be proved to be false—human evolution is a story of successful and unsuccessful trying—but it may be that it will be proved to be one of the most important evolutionary hypotheses which human beings have ever thought up for themselves.

The Principle proposes a quite different way of living and of seeing one's life; not different in the sense of making its users into oddities but different in that its users can learn to adopt a different yardstick for themselves and for the people they live with. Different in that its users—over the thirty years I have observed it—seem to be able to adapt more successfully than most people in their social, artistic, and biological spheres: and, most important of all, they appear to live longer and more healthily.

The Alexander Principle states that there are ways of using your body which are better than certain other ways. That when you lose these better ways of using your body, your functioning will begin to suffer—in some important respects. That it is useful to assess other people by the way they use themselves. That however clever or powerful a man may seem to be, however well-endowed a woman may be with beauty and charm, however rich people may be in cash and in contacts, they are suspect if their USE is suspect.

This approach is not a fringe-medicine, a neo-progressive education, a religious escape, or a quack science. It is a difficult disciplined approach to personal living which leads, through discipline, to a personal freedom and health which is possible to some extent for most people at most ages.

There are few people who would not find something useful in what Alexander had to say. He had much to say to the individual, and he had much to say to the educational profession: but he was also laying the foundations for a whole new science of USE, and he was opening up a whole new field for medical enquiry.

Chapters 2, 3 and 4 will discuss his basic principles of USE, BALANCE and REST and to these we must now turn.

Chapter 2

USE

MOST OF US are fatalistic about our bodies. According to the luck of the draw, we expect to grow up tall, short, plump, thin, weak, muscular, graceful or clumsy. We expect when we are young that we will grow up and grow old, and we expect that as we grow to middle-age, we will deteriorate. We think that our structural faults lie in our stars and in our parents, not in ourselves: that our body-potential is immutably limited by our initial genetic programme.

To some extent we are right, but to say this is not to have said much more than that the game of chess is boringly limited by the black and white squares.

The Alexander Principle insists that our will is potentially free: that it is what we do with our genetic inheritance which will determine our future structure and performance in nearly all the ways that matter. It starts at a relatively crude level by drawing attention to the ways in which we use and misuse our bodies in such simple matters as standing, sitting, and lying down, and it says that even at this crude level there is USE which is beneficial and USE which is harmful. It goes on from this relatively crude observation of mechanical mis-use to show how the basic structure of the personality, at its minutest and most intimate level, is fashioned from our BODY-USE.

William Harvey, in 1616, described the circulation of the blood, and thereby revolutionised medical thought. This did not mean that prior to this time the blood had not circulated and that it suddenly started circulating there and then; it had circulated for aeons of time before Harvey first described, albeit imperfectly, what was going on. In the same way, the type of USE which Alexander described, has been present for aeons of time. Before Alexander, much was known about it but not in a way which could helpfully be applied to man's health.

It is a long way from William Harvey to Christiaan Barnard. What is written here will no doubt seem elementary in a hundred years' time, and indeed, since Alexander's death, his procedures have been constantly refined; no doubt many more false leads will be attempted and have to be abandoned. It has to be stated clearly that the type of USE which I describe in this book cannot be considered as THE ONLY RIGHT WAY: the USE which I describe is the best I have been able to discover, and, as I describe it, it works. But no doubt far better ways will be found eventually of describing and refining this new approach.

Alexander's discovery of Wrong Use

Alexander was born in 1868, in rural Australia. To judge by his reviews, he was a successful young actor, in Sydney, until he was increasingly plagued by voice trouble. In the nineteenth century little was known of speech training or speech therapy as we now know it, and Alexander's recurrent loss of voice brought his stage career to an untimely close.

In desperation, and with little medical or physiological knowledge, he decided that he must observe minutely, in a mirror, the way he was using his muscles when he spoke. It is a common observation that when people speak, they are liable to carry out quite inappropriate movements throughout the whole of their body: a glance at a television screen will often show announcers and commentators who have persistent mannerisms when they speak. Alexander was particularly struck by curious movements which took place around his neck and head as he spoke: and although the types of curious movement which can take place in this region are numerous, he picked out the most common one, which consists of a tightening of the head backwards on the neck and downwards into the chest.

At this point—and indeed for the remainder of his life—he became concerned with the types of muscular usage which arise when people react to a stimulus. He was, in fact, a child of his times, with its Stimulus-Response psychology and the Behaviourism which Pavlov and his dogs helped to foster. Fortunately Alexander's initial observations provided enough impetus to enable him to develop and refine his methods until the end of his days. It is necessary to understand—and not to be daunted by—the exuberance which he showed about his initial discovery: a reasonable exuberance, since his voice problems cleared up when he had learned how to stop pulling his head back and down.

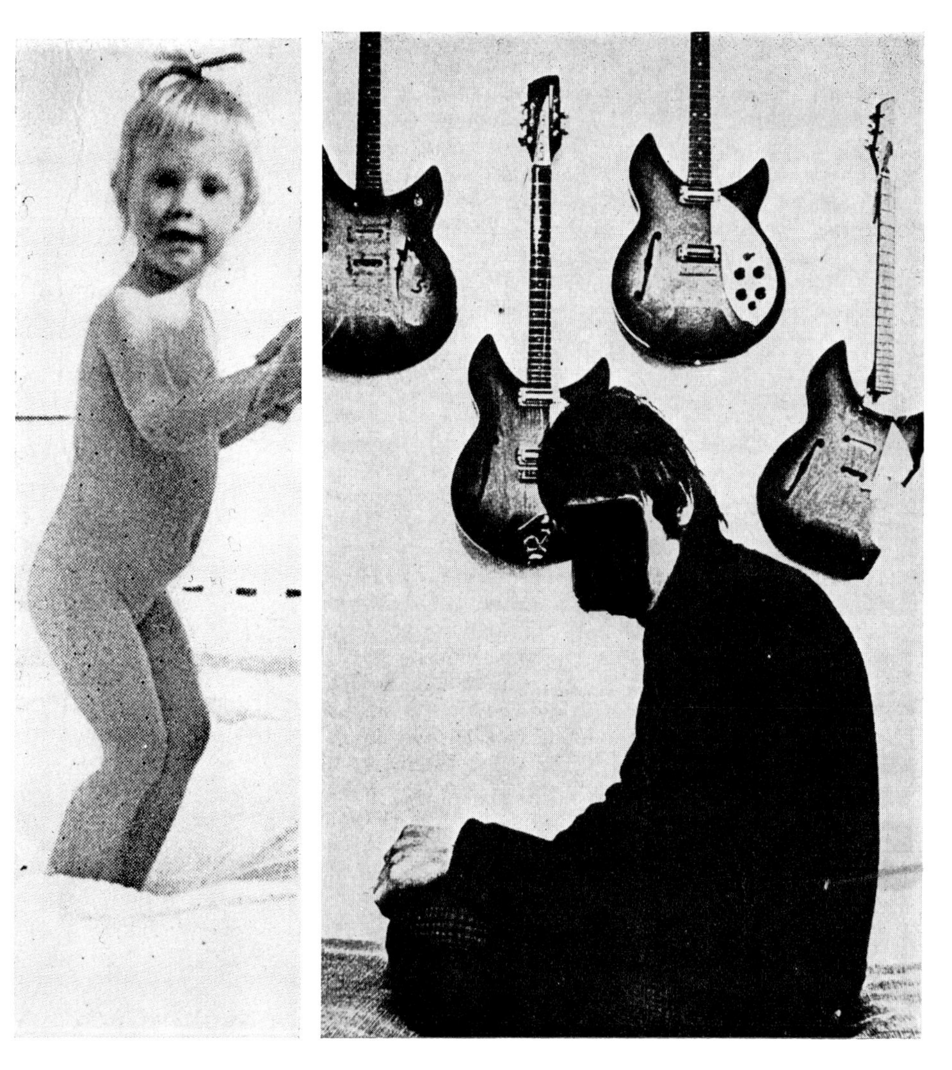

3a 3b Infant alertness versus adolescent collapse.

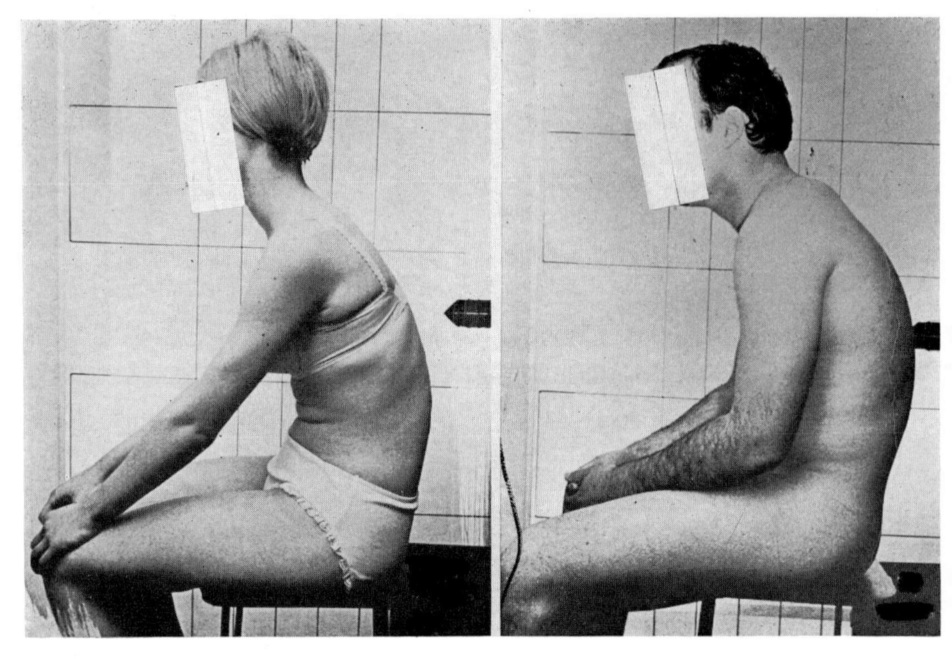

4 Typical slumped sitting positions.

5 Two types of head retraction.

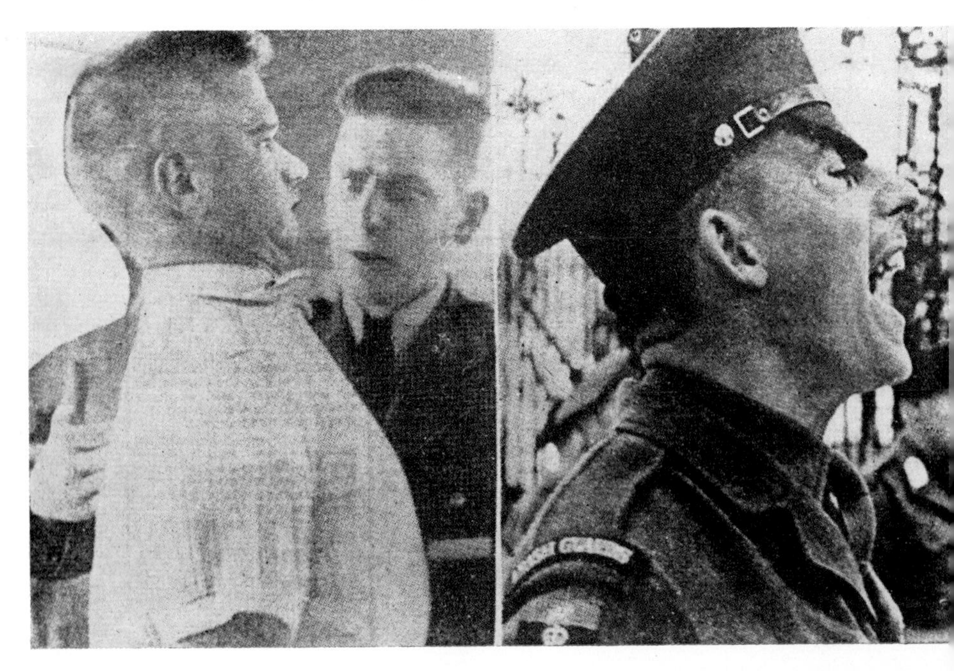

The Primary Control

Alexander's observation of his unconscious mis-use of the neck and head led him to term his improved USE "The Primary Control". He wrote in *The Lancet*: "When I was experimenting with various ways of using myself in an attempt to improve the functioning of my vocal organs, I discovered that a certain use of the head in relation to the neck and of the neck in relation to the torso . . . constituted a '*Primary Control*' of the mechanisms as a whole."

Alexander and some of his supporters at one time seemed to impute an almost magical significance to the "Primary Control" and some of his medical friends gave him information about "controlling centres" in the mid-brain in terms which seemed to imply a subjective awareness of such a centre, which could exert a "Primary Control" over the rest of the body. Shades of Descartes and his Pineal Body!

Few people would find it helpful nowadays to talk about a "Primary Control", although in the past the phrase did emphasise the prime importance of a proper USE of the head and neck, at a time when anatomists and physiologists had no very clear account to give of the factors underlying balance. And fortunately the "Primary Control" hypothesis did not restrict the development of Alexander's practical teaching methods, although it certainly affected the emphasis of his teaching.

Head Retraction

Alexander wrote further of his observations as follows:[1]

> If you ask someone to sit down, you will observe, if you watch their actions closely, that there is an alteration in the position of the head, which is thrown back, whilst the neck is stiffened and shortened.

This is as good a place as any to begin and soon after I became interested in Alexander, I thought I would see if what he said about this was in fact true. I had the opportunity to carry out an experiment with 105 young men, aged between 17 and 22. It is an experiment which can be tested by anyone. Plate 1 (facing page 12) shows a young man and a girl sitting down. It can be seen that both of them are throwing their heads backwards in the process. When I tested the 105 young men, I fixed a tape-measure to the back of their heads, and made an ink mark over the prominent vertebra where the neck joins the chest at the back (fig. 1a). I then asked them to sit down

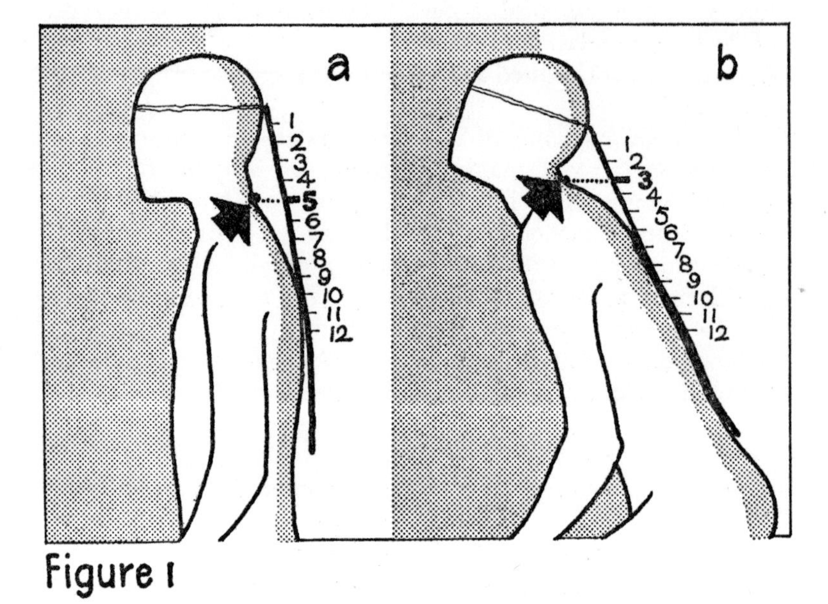

Figure 1

and whilst they were sitting down I observed how much the tape-measure moved down over the ink mark (fig. 1b).

Out of the 105, only one did not pull back his head and move the tape downwards; 56 moved it down two inches or more, 43 between one and two inches, and 9 under an inch. The younger ones pulled their heads back-and-down less than the older ones. What was even more interesting was that when I asked them to prevent this, only 11 of them were able to stop contracting their heads into their shoulders, however hard they tried.

Sir Charles Sherrington, the neurophysiologist, who drew encouraging attention to Alexander's work in the 1930's, pointed out what a complicated thing it is to do even such a simple thing as sitting down:[2]

To execute it must require the right degree of action of a great many muscles and nerves, some hundreds to thousands of nerve fibres, and perhaps a hundred times as many muscle fibres. Various parts of my brain are involved in the co-ordinative management of this, and in so doing, my brain's rightness of action rests on receiving and despatching thousands of nerve messages, and on registering and adjusting pressures and tensions from various parts of me.

It is not surprising that most of these young men could not alter their habitual head-movement simply by trying not to. We cannot alter our habitual way of doing things simply by deciding to do them some other way. Our will is potentially free, but to free it for effective action, we need certain principles of USE on which to base our actions.

The Use of the Head

Of the 105 young men, 104 showed this particular use of the head in which the skull was pulled back on the neck. The first thing to decide is whether such a habitual use of the head and neck is important. Let us simply say that according to the Alexander principle plate 2a is a MIS-USE. In the last analysis, it will be a matter of taste whether people prefer some other forms of USE to Alexander-USE: indeed whether they would prefer to be sick rather than healthy, to be in pain rather than comfortable.

To me it is self-evident that the state of the child in plate 2c is better than in plate 2b: that plate 14c looks better than plate 14a: that something is amiss if children emerge from adolescence looking like 3b: that the slumped appearance of the two students in plate 4 is aesthetically unpleasing and harmful, amongst other things, to their ability to breathe. Yet such states of slump and head retraction are nowadays the rule, in homes, in classrooms, and in lecture halls; in theatres, in cars and public transport, and, of course, in mental institutions and parliament; and in church as well.

The Sedentary Life

Few of us will spend more than an hour or so at a time without sitting down. Many people spend more of their working day sitting down than moving, except of course for young children who, once they can walk, can only be made to sit still for a short time by being restrained in high chairs and seat harnesses, or by their mother's hands and arms. The young child instinctively moves and explores and communicates as soon as it wakes and it will continue to do this until it is tired, or until it has been rebuked or restricted into a stillness which is socially more convenient. Such restriction is an inevitable feature of school life, and eventually the growing child may well be sitting at the desk for hours on end, in a class with perhaps forty other children. Most of the children in the classroom will be sitting in a collapsed state, with the weight of the trunk supported through the elbows and the shoulders. The sedentary life has begun.

If then, so much of our life is spent in a sedentary position—and most of the "top" men and women in government, industry medicine, law, music and education and so on will have spent many years sitting down whilst they equipped themselves for their job—it would seem important to consider how we use ourselves in that position.

Sitting Down

How do we usually set about getting ourselves "sat-down"? Over 99% of us, as exemplified by the 105 young men—and confirmed by studies on all age groups from puberty upwards—pull the back of the skull down into the back of the neck as we sit down and stand up. Usually—unless we are actors or dancers—we are relatively unaware of how we use our bodies as we carry out our daily activities. When we want to sit down, we walk to the available seat, rapidly gauge the seat height, and plant our bottoms without further ado, in such a way as to avoid other people and other objects and without showing too much thigh. In the process, the head usually pulls back and the spine becomes curved. When the seat has been reached there are a few shuffles and wriggles to eliminate the creeping and crinkling of clothing and then the body is allowed to collapse whilst the head and neck are kept in a position which will allow social intercourse or reading or writing. This more often than not involves using the arms and shoulders as struts to support the collapsed body. For eating, the face drops down towards the plate. For tele-viewing, once the initial hypnosis has been induced, the body is collapsed to the lowest point of slump at which the eyes can look ahead out of their sockets. The miracle is that human beings survive it at all: the tragedy is that they know no better and that by the time their body begins to cry out "Enough, Enough", they are set in their ways and in their social commitments.

Neck Collapse

Sitting-down has been mentioned first because so much of our time is spent using ourselves in this position, and also because the mis-use in this position is so obvious. When Alexander studied his head movements and positions during speech, the mis-use was not so obvious, although plates 5a and 5b show

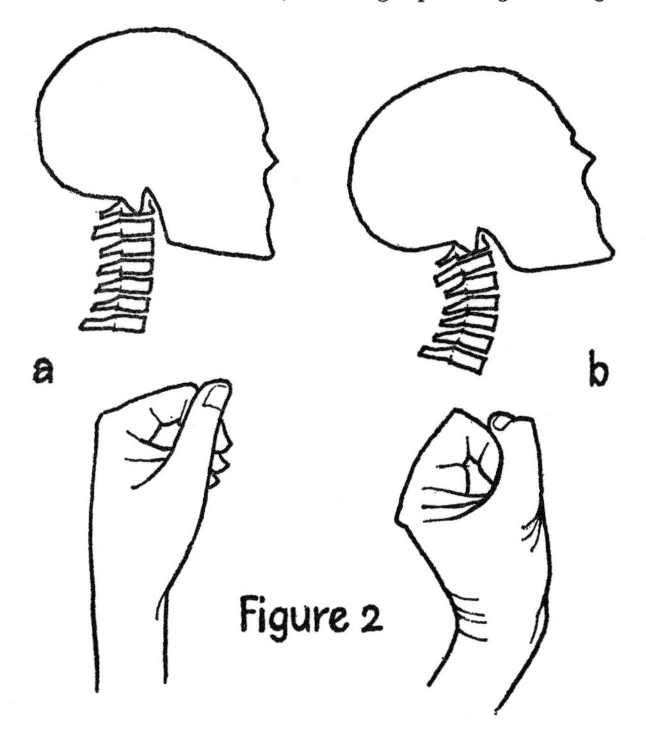

Figure 2

only too clearly how gross the wrong movements of the head can be. And if we consider the picture of a dentist in plate 6 and observe his head and neck position as he bends over a patient, it is clear that his mis-use consists of a dropping of his neck forwards, and a collapse of his upper back into a curve. The movement can be illustrated by the hand and wrist (fig. 2a). The dropping forward of the wrist corresponds to the dropping forward of the neck, and this is shown on a skull in figure 2b. They can be perhaps more clearly seen in X-rays of the neck, plates 7a and 7b. There are a myriad nuances of possible muscular mis-use, but these most common and obvious mis-uses need first to be emphasised: pulling the head back and down is almost universal in our culture.

What follows from this collapse of the neck and upper back? It means the gradual development of a persistent HUMP at the base of the neck, and it means that the rest of the body, if it is to balance itself, has to be wrongly compensated elsewhere. Thus the collapsed neck leads to a throwing-in of the back—the lordosis which is so obvious in some coloured people (plate 8) and which remedial gymnasts try in vain to correct in school children (plate 9a–9b). The almost pandemic complaint of low back pain (ranging from housewives' lumbago to the outright slipped disc) cannot be adequately treated by simply concentrating on the lower back. In by far the majority of cases, the lower back deformity is consequent on a deformity in the Hump—consequent, that is to say, in the sense that only when the mis-use in the upper back is corrected can there be an efficient stabilising of the lower back. (See also fig. 22f.)

When Alexander first noticed the mis-use of his head and neck, he was noticing something of far greater importance than simply something which affected his speech. He was noticing a defect in the most fundamental structuring of the body, and he made the correction of this defect a fundamental part of his principle that USE affects Functioning.

What is a Principle?

Since this book is about the Alexander *Principle*, a statement here by Gilbert Ryle about what we mean by calling something a principle may be helpful.[3] Gilbert Ryle says: "A question is a question of principle when it is much more important than most other questions. The relative importance of questions can be explained on these lines: that when, given the answer to one question, it is at once clear what the answers to an *expanding* range of questions are (whilst the answers to any of these other questions do not in the same way throw light on the first question) then the first question is a *question of principle* relative to the latter. Or when, in the case of a range of questions, it is clear that none of them could be answered, or perhaps, even be clearly formulated, before some anterior question is answered, then this anterior question is a question of principle relative to them. The notion of a principle is simply that of one question being logically prior or cardinal to a range of other questions. It is tempting, but it would be too rash to say that there is one absolutely first question."

I have quoted Ryle at length, since it bears on the primacy of Alexander's principle of "The Primary Control". The mis-use of the head and neck is prior to mis-use elsewhere, according to his principle. Mis-use elsewhere can only be adequately dealt with after the correction of mis-use of the head and neck. To paraphrase Ryle, "When, in the case of a range of problems of bodily mis-use, it is clear that none of them can be dealt with or, perhaps, even clearly formulated, before some anterior problem is dealt with, then the need to solve this anterior problem can be termed a principle."

It will be clear from this that we have to start by asking the *right* questions about our body: and we have to find the answers to an expanding range of questions which are thereby thrown up. Alexander's observation of mis-use during speaking opened up just such a new and expanding range of questions.

The Hump

Most parts of the body have two sorts of names—an everyday common-usage name and an anatomical name (or group of names). For example the names "Elbow" or "Wrist" or "Shoulder" are understood by us all, although in fact they are not very clearly defined anatomical areas: the wrist is, for most of us, a vague area between the hand and the forearm which is liable to get strained: the shoulder is sometimes thought of as the shoulder-blade, or as the top of the upper arm, or as the shoulder-joint itself. The same thing applies to most common body terms—the knee, the stomach, the chest, the back, the neck, and so on—and for most purposes we get on fairly well with these vague terms and we would be lost without them.

The astonishing thing is that the most important USE-area we possess, from the Alexander point of view, has no common name. My task of describing USE would be greatly simplified if there were a simple commonly accepted name for the area of the body in which Alexander first detected a fundamental mis-use, an area which is indeed the Clapham Junction of USE.

It is the region of the body where the back of the neck joins the upper back, and it is a prominent area in everyone because here the shape of the neck vertebrae alters and the spinous processes become more prominent. Figure 3 shows it in profile and plates 10a, 10b and 10c show it in a young child, a lady, and a middle-aged man.

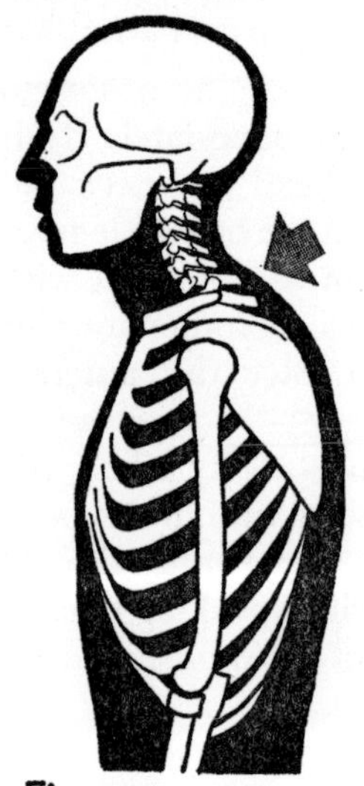

Figure 3

This whole region at the base of the neck, both back and front, is a veritable maelstrom of muscular co-ordination. It is here that those most inadequate evolutionary adaptations—the shoulders and upper arms—will exert their distorting influence during the many activities in which we engage. It is here that faulty patterns of breathing throw the muscles of the lower neck and upper ribs into excessive spasm: it is here that mechanisms of speech and swallowing require a reasonably good vertebral posture if the oesophagus and trachea and associated vocal structures are to function well. It is close to here that blood vessels and nerves of great importance and complexity will pass—blood vessels to the base of the brain, nerve ganglia which affect breathing and heart rate and blood pressure, nerve roots which with increasing age become more and more liable to compression: it is here that 85% of the readers of this book will have arthritis by the time they are 55 (and many of them much younger than that): and it is from here that the head itself—the structure which carries man's most important sensory equipment of sight and hearing, taste and smell, and balance—has to be co-ordinated at rest and in movement.

And it is here that mis-use most frequently starts: it is here that we have to start if we are to correct the multitudinous mis-uses which the rest of the body can throw up. In terms of the Alexander Principle, it is only when this primary mis-use is dealt with that we shall see the answers to an expanding range of questions.

In the past, the hump has been thought of as a dull, inert, fleshy region, with little of interest to offer except to the painter or shot-putter or dowager pearl-wearer: and indeed, when improved use has been established, the hump-region simply provides a context within which other functioning can take place.

What produces the hump? The short answer is excessive and wrongly distributed muscle tension. By habitually moving and keeping still in certain ways, we gradually alter our physique. Our manner of USE, at rest and during movement, contains a tangible record of all the basic habits which we have laid down over the years. In most people, the HUMP is a tangible witness to a lifetime of mis-use.

The phenomenon of head retraction which Alexander first noticed is a symptom of pre-existent muscle tension, not the cause of it. Alexander, with simple clarity, proposed that if only people could stop pulling their heads back whenever they reacted, all would be well, and he concentrated his efforts on training both himself and his pupils to stop doing just this. Fortunately for him, there was enough novelty and truth in this fact to help himself, his teachers and his pupils for a whole life-time: there was enough in it to improve very considerably the functioning of those who understood the discipline: there was enough in it to infuriate some members of the medical, educational and other learned professions, since they could not decide what he was getting up to—H. G. Wells rudely called it "swanning", others called it quackery and were astonished that men of the calibre of Professor Dewey, Professor Raymond Dart, Archbishop Temple, Stafford Cripps, and large numbers of doctors could be such enthusiastic supporters.

It was inevitable—and still is inevitable—that, once some of the main features of Alexander-USE were described, people would find themselves boringly lost in cul-de-sacs which had to be explored before further roads could be taken. This early phase of Alexander's work was an inevitable progression from the work of the early anatomists and anthropologists, who were concerned with balance and the upright posture, and we must turn our attention in the next chapter to this question of balance and posture.

Chapter 3

BALANCE

AMONGST THE BOSOMS and bottoms of seaside-front picture postcards, there used to be one of a decrepit old man, standing unsteadily in a doctor's surgery, legs splayed out, holding on to the furniture and saying, "Well, doctor, how do I stand?", and receiving the inevitable answer, "Honestly, I can't imagine". When I look at the hunched backs, twisted spines, fixed pelvises and hopelessly inadequate legs and feet which trudge through my clinic, I also often find it hard to imagine just how they manage to stay upright at all!

The fact is that, for most people, balance is not a question of tight-rope perfection, of ski-jumping precision, of the pas-de-deux or the coolness of the mountaineer. It is possible to sit, stand and walk and indeed to perform highly skilled tasks and yet to be wrongly balanced. The skater in plate 11 can do something which most people would find impossible. She earns her living by bending double and then skating backwards in order to pick up a handkerchief from the ice with her teeth. Obviously she has a keenly developed co-ordination for this and similar activities, but yet her balanced USE of herself is wrong.

Close analysis of the photographs shows that her rib cage is twisted over to one side, not only when she bends but when she is standing still. Had it not been for the fact that she eventually developed pain in her back whilst skating, she would have been totally unaware of this imbalance in her back. Indeed her dance teacher, her doctor and an orthopaedic specialist had not noticed the twist, and she was unable to stop her pain until she learned how to use her back with a more symmetrical balance. Like most people, the dancer took it for granted that her body was a reliable instrument, that it had its own unconscious "wisdom", and that as long as she could do the work for which she was trained, her mode of balance was adequate.

Body Wisdom

One of the legacies of the last century, with its accent on the God-given perfection of the human frame, was what W. B.

Cannon later called the "Wisdom of the Body".[4] Cannon suggested that there are certain balanced states of the body which are natural and normal and to which, in its wisdom, the body will return after disturbance and stress. Such "body wisdom" was taken to apply not only to muscular balance, but to the organic constituents of the body. Illness, on this analysis, is accompanied by states of imbalance—the sugar in the blood is constantly raised, the bowel constantly overfull, the vital capacity of the lungs constantly diminished. On this view, the physiological wisdom of the body has to be restored by appropriate medical treatment and care, until a more normal resting balance can be maintained—with or without drugs. More and more drugs begin to be needed to keep the blood pressure down, the heart beat regular, the sleep pattern tolerable. A mental "resting-state of balance" is likewise to be achieved by more and more drugs to stop anxiety and depression, or else by the cultivation of a Nirvana state by meditation or other spiritual disciplines.

It is now clear that the "Wisdom-of-the-Body" story is a fallacy. Increasing dependence on therapeutic drugs—however wisely and cleverly they may be prescribed—is proof that most people's body wisdom has gone astray. Somewhere along the line, the whole story of physiological wisdom has gone off the rails.

Nowhere is this more clear than in the muscular balancing mechanisms which underlie USE. In recent years, deservedly popular books like Desmond Morris's *The Naked Ape*[5] have made people aware of the mechanical problems which the upright posture produces. Many nineteenth-century anatomists had assigned varying importance to the upright posture and its accompanying blessings. Amongst some of them, it was customary to see something partially divine in being upright—"that majestic attitude which announces man's superiority over all the inhabitants of the globe". The Naked Angel, in fact, rather than The Naked Ape.

In the first quarter of this century, it was still thought that our spines were perfectly fashioned for the upright posture, but that the world we lived in was to blame. Sir Arthur Keith,[6] who was the authority on posture in the 1920's, thought that postural defects were caused by "the monotonous and trying positions which are entailed by modern education and modern industry".

But in the second quarter of this century, the new speciality of orthopaedics came increasingly to the view that it was not man's environment but his imperfect adaptation to it that was at fault. Man was seen more and more as a made-over animal, with muscles forced by the adoption of unnatural stances to suffer enormous inequalities in the distribution of labour.[7]

In the second half of this century, a new speciality called Ergonomics turned its gaze on how to fit machines to man and man to the machine. Chairs, car seats, beds, desks, and all sorts of complex machinery have been designed to give correct elbow room and leg length, to give better positioning of pedals, levers and display panels and proper seat dimensions. It was hoped thereby to minimise the fatigue and strain of unnecessary movement in faulty positions.

However, owing to the lack of an adequate concept of muscular balance, the ergonomic approach has not paid off the expected dividends. The working man still arrives home fatigued. Pain in the back affects most of the population, often with crippling severity. 75% of our dentists develop troublesome back pain, over 80% of our secretaries develop headaches —they have not been helped very much by better designed equipment. It is their USE which needs redesigning.

Posture both in the home and in the work place, is as poor as ever it was—and indeed, my figures for adolescents show a deterioration over the past twenty years. Even in the most perfect environment which the ergonomists can construct, the "Wisdom-of-the-Body" cannot apparently resolve the conflict between those parts of the body which are needed for doing an actual job, and those parts which are needed to support the general functioning of the body. Mis-use, in our time, persists and increases.

Living Anatomy

This lack of an adequate concept of muscle balance came originally from the dissecting-room. Anatomy, as studied in the corpse, does not bring out the complexity of muscular mis-use in the living man.

The identical sameness of the muscles of any man and woman can indeed be demonstrated in the dissecting room without a shadow of doubt, and students can learn in great detail the

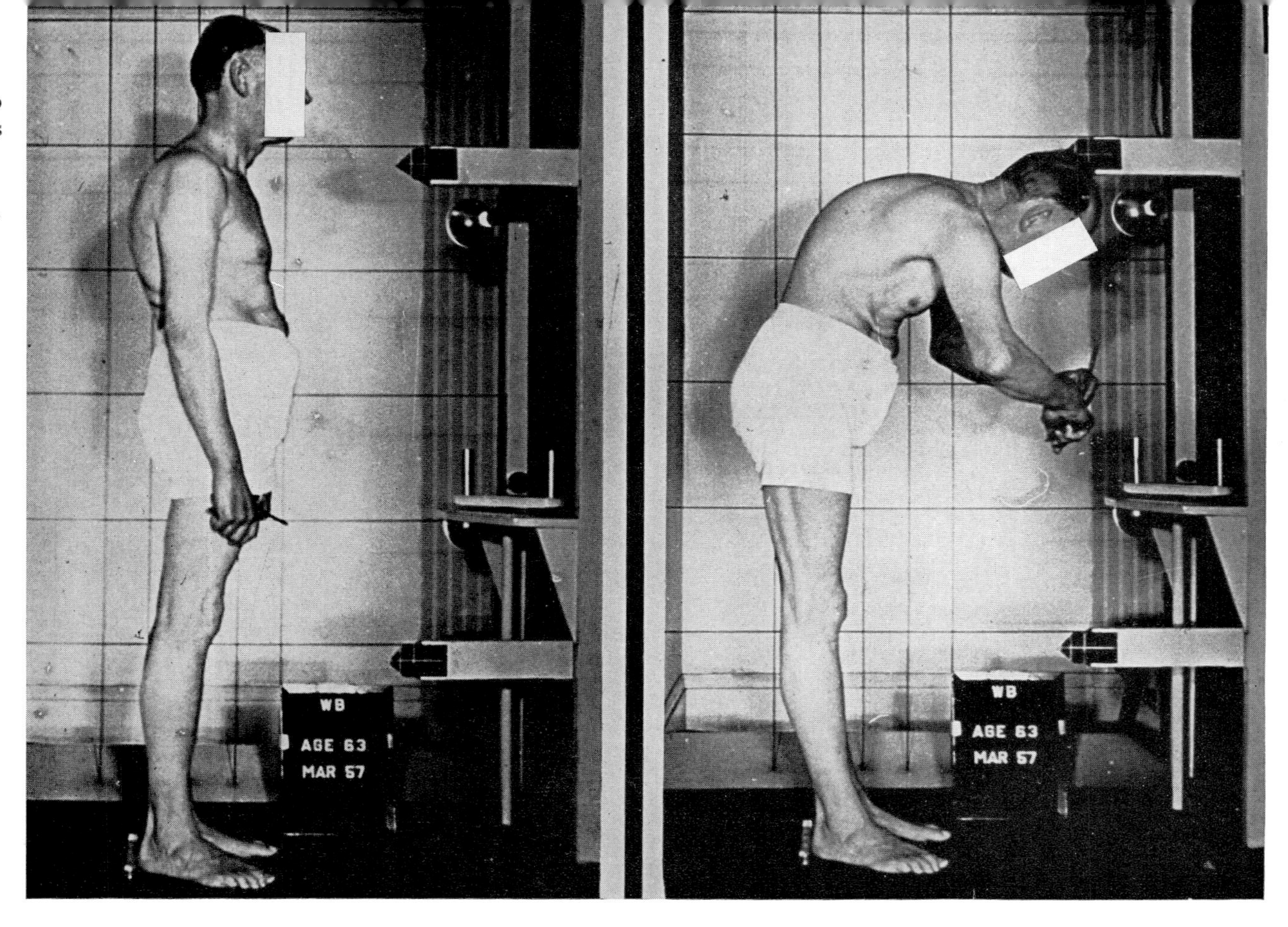

6 Dentist's hump produced by years of bending over patients, and pulling head back.

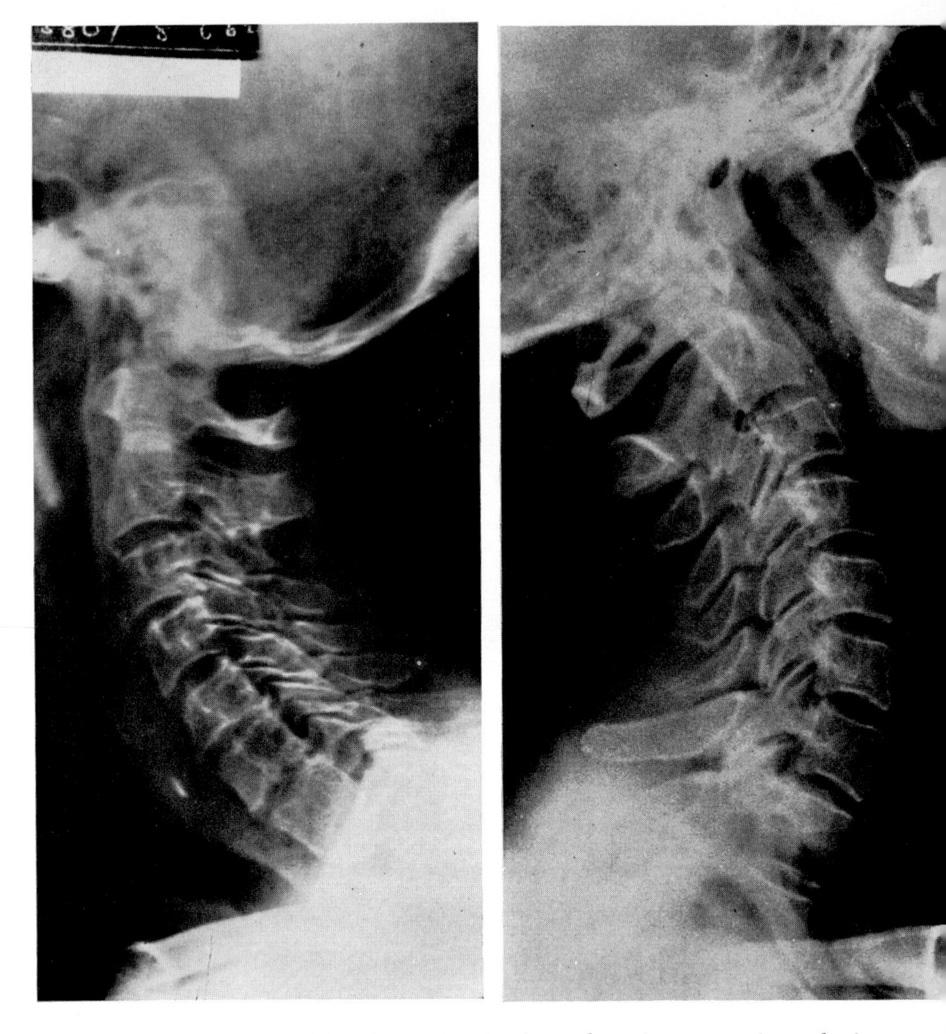

7a 7b X-ray of heads pulling back, and neck constantly arched too much forward as in Plate 2a.

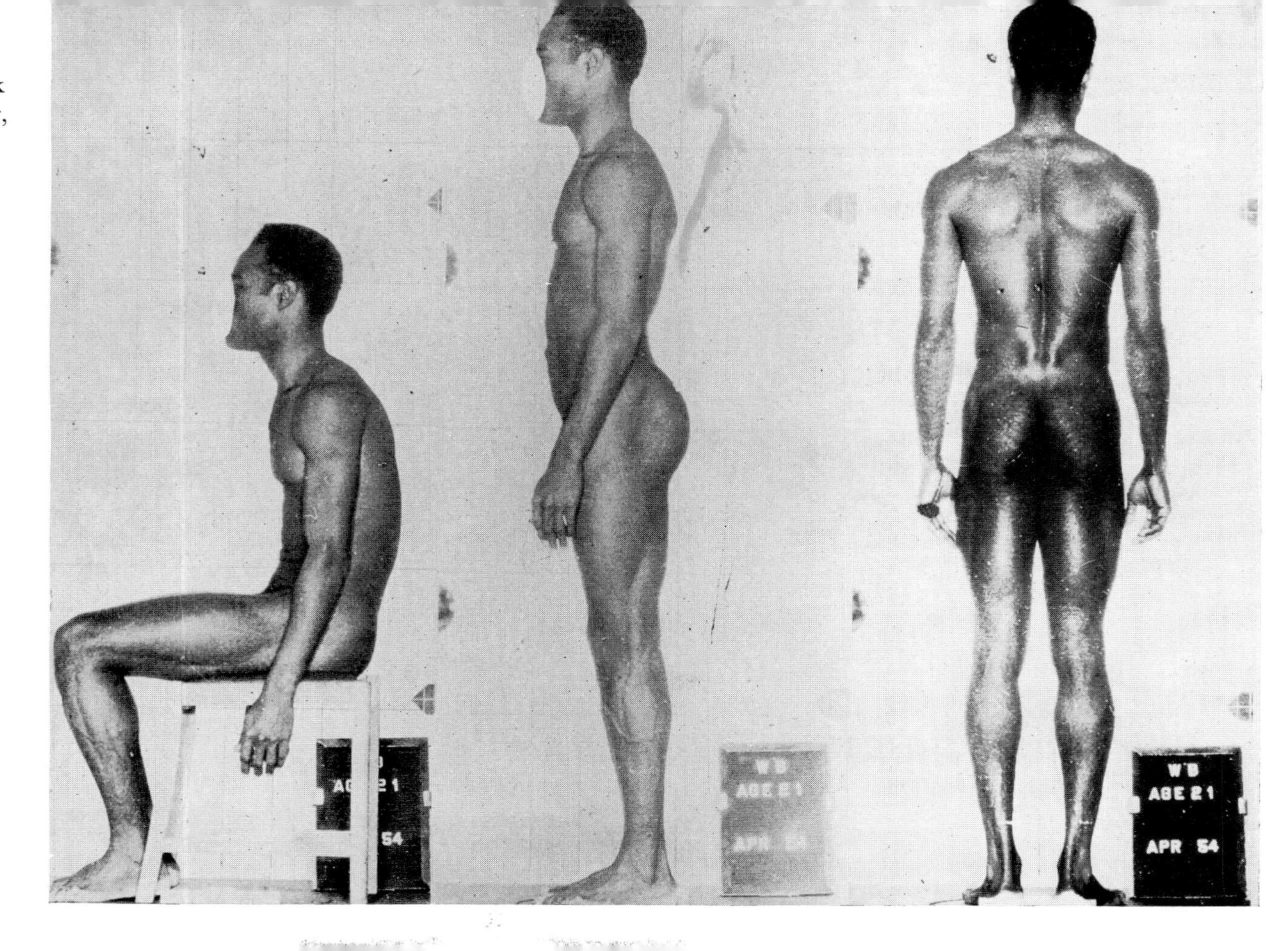

8 Athlete with back slumped when sitting, over-arched when standing: excessive tension in back muscles.

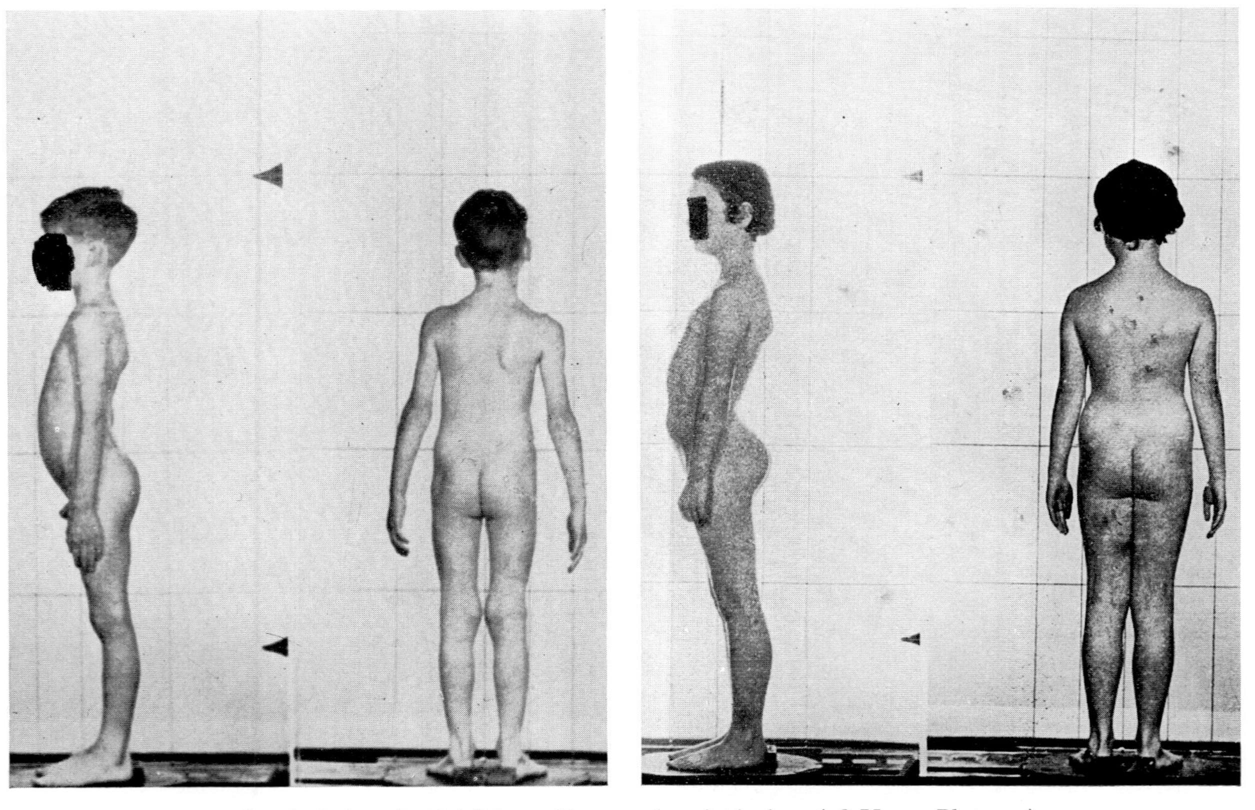

9 Lordosis in schoolchildren. Note neck twist in boy (cf. X-ray Plate 22).

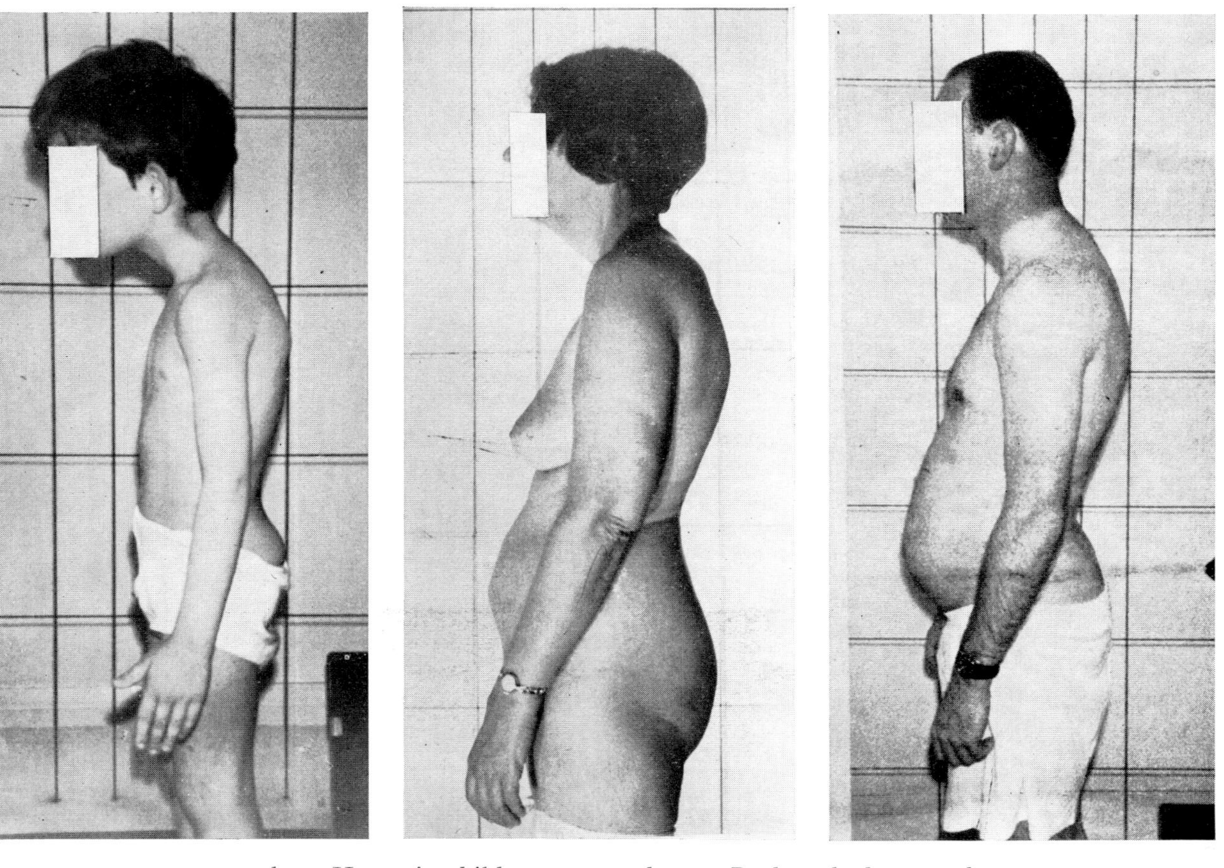

10a, b c Hump in child, woman and man. Back arched, stomach dropped forward. See fig. 5 and fig. 22e.

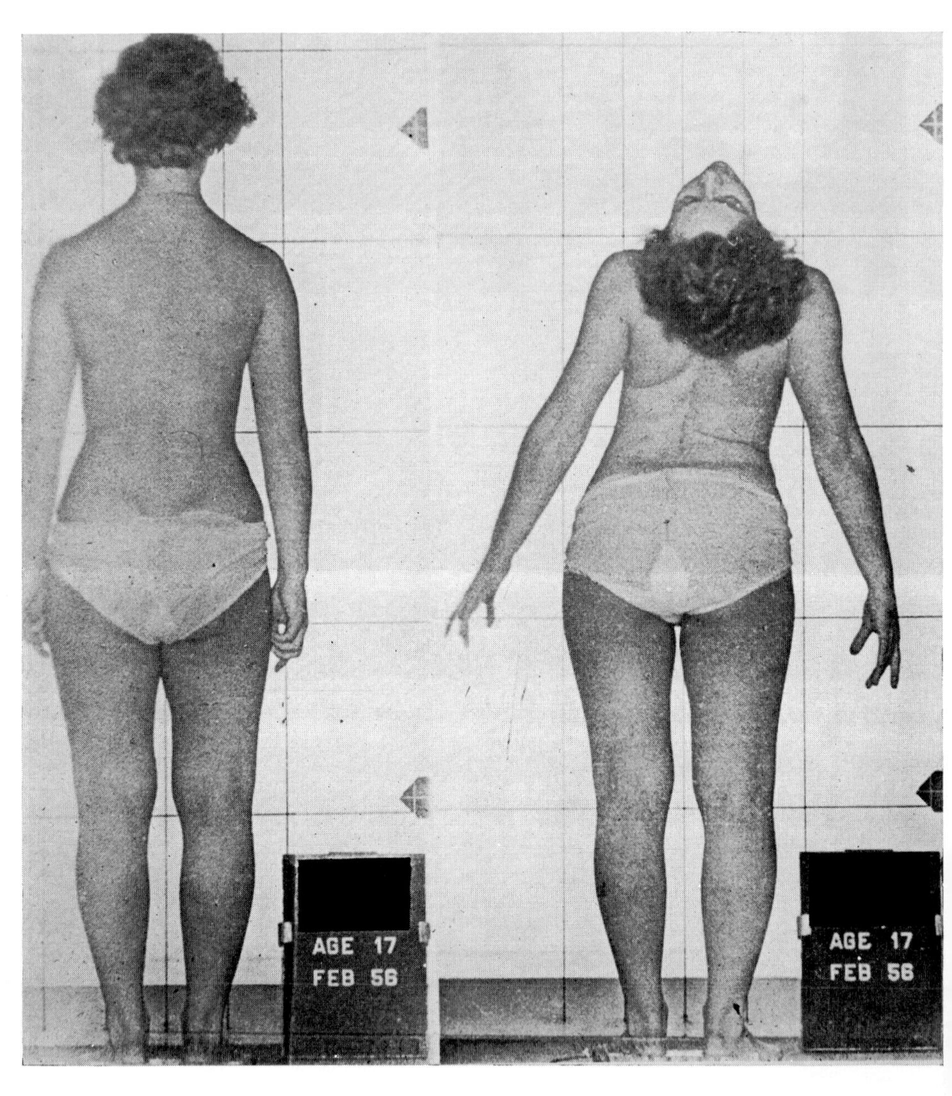

11 Dancer unconsciously twisting her back when bending.

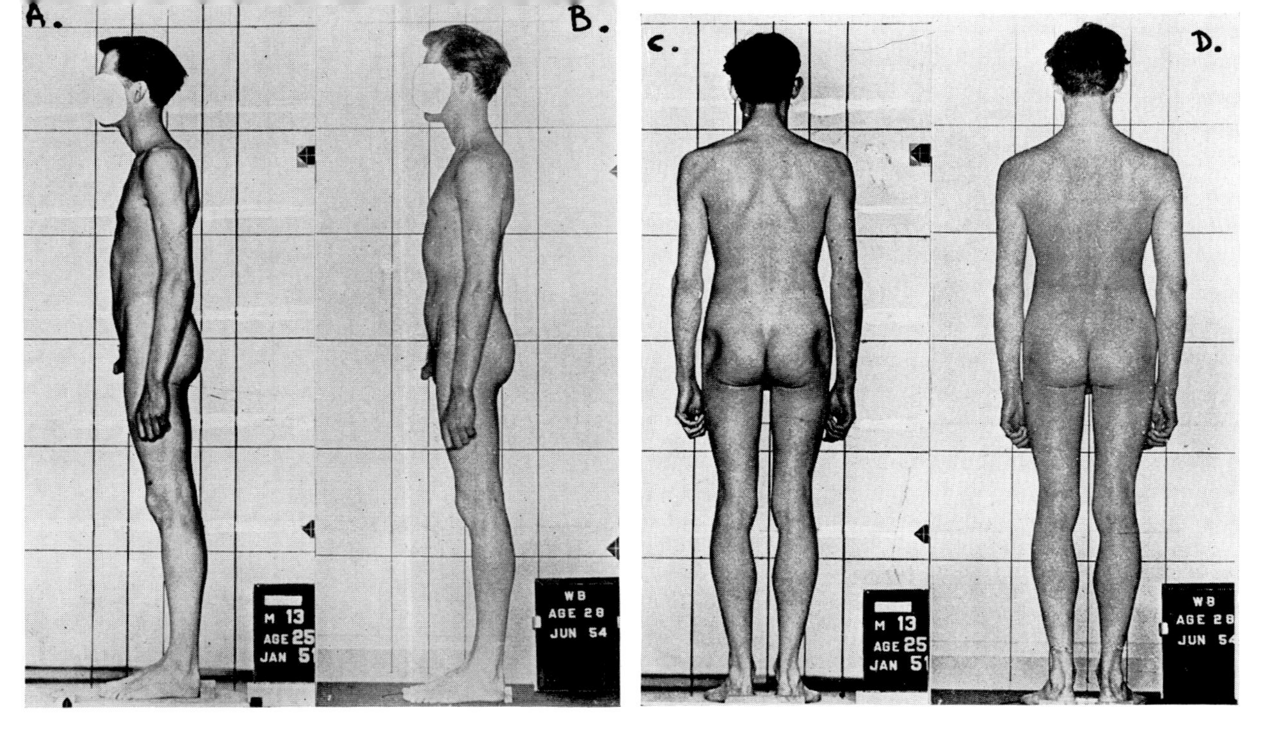

12 a. Neck dropped forward as in X-ray Plate 7.
b. Neck lengthening as in figure 5b.
c. Back view of 12a. Note tension in neck, shoulders and buttocks.
d. Back view of 12b. Less tension in neck, shoulders and buttocks.

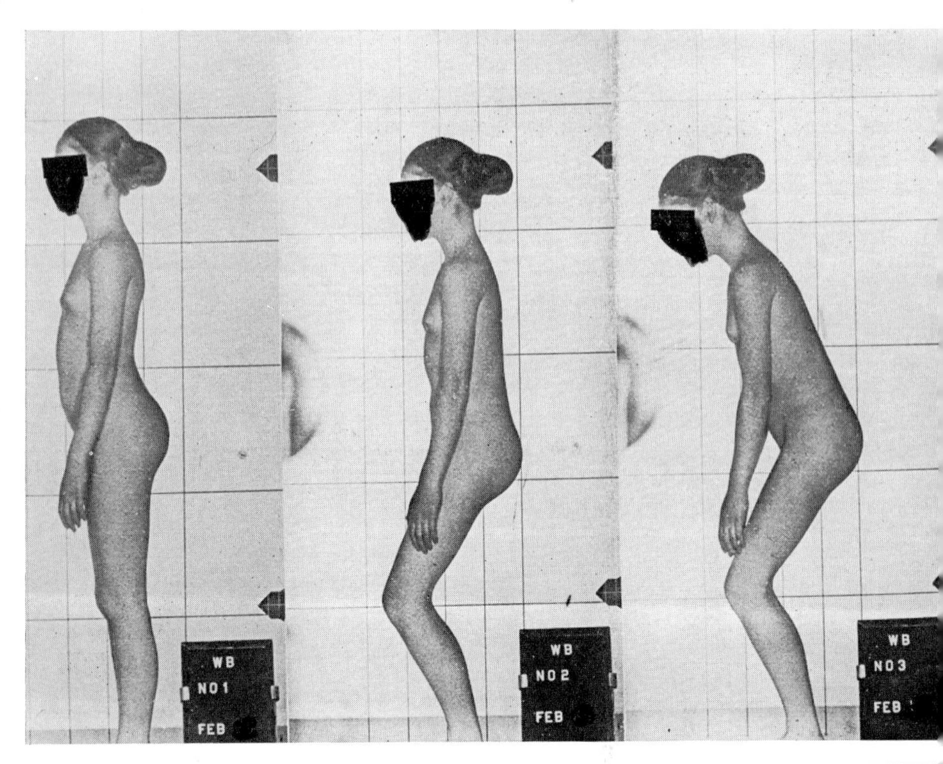

13a Alexander's "position of mechanical advantage".

13b (*right*) Head retraction into position of mechanical disadvantage in sitting down.

names and actions of hundreds of muscles which pull on various body levers throughout the body. But this is the anatomy of the dead, the anatomy of sameness.

A living anatomy has to start with the living body and with the infinite variety of each of us. To give them their due, the ergonomists have always understood the need to study muscle action in the living: but unfortunately they took *mis-used* man as their norm, and much of the equipment which they designed was designed for mis-used man. Only rarely did such equipment encourage really good use, and then only in a most superficial sense. The typist may sit in the statistically perfect chair, but her basic habits of mis-use still persist.

The early anatomists and ergonomists have set the stage for a real understanding of the problem. We now know that man's body is not majestically divine. The "Naked Ape" has replaced Rousseau's "Noble Savage" and Wordsworth's "Nature's Priest". We now know that we are faced with an evolutionary problem of combining the potentiality of an intelligent angel with the impulses of an irascible ape.

Alexander, born into the last quarter of the nineteenth century, was plunged straight into this evolutionary argument. When Darwin and Huxley made their onslaught on the first book of Genesis, their evolutionary man had still to compete with the image of the divinely created man. He could not be allowed the simple animality of the ape, but he had to have a splendour and grace of his own: his body had to have a natural goodness and wisdom. In this way, Wordsworth's "Nature's Priest" (who by the vision splendid, was on his way attended), could hold his head high in competition with the divine image.

It is interesting that Alexander's first book was entitled *Man's Supreme Inheritance*. He was inevitably caught up with the notion of a basic perfection which is lost by a combination of environmental stress and personal stupidity. His theme during his lifetime was of an endowment of properly functioning reflexes, which the corporeal sin of mis-use (induced by the overstimulating newness of the environment) has clouded over. On this view, a system of perfectly adequate reflexes had to be restored by learning to inhibit wrongly acquired conditioned reflexes. Or in the words of a recent cleric, by stopping off "ugly contradictions in the true nature of man".

The "vision splendid" has now no need to postulate the

perfect God-given or gene-given templet—one correct shape, and one only which is appropriate for our human stance. We have no "true nature", beset with "ugly contradictions". To know and to be what we "truly" are, we have to *find out* what we are; and we have to construct what we are to be.

The Alexander Principle suggests that by getting our USE in the right order, there is a chance of a new personal evolution. There is no reason to suppose that we are born with a perfectly ordered set of pre-existent natural reflex patterns, and that, by refraining from interfering with them, all will be as well as it can be. The next step in our evolution has to be learned by each one of us, ourselves. *Personal selection has to replace natural selection.* Instead of relying on what the environment can winkle out from our possibilities by natural selection, it becomes a question of what *we* can winkle out of our possibilities by personal selection.

What sort of balance should we personally select?

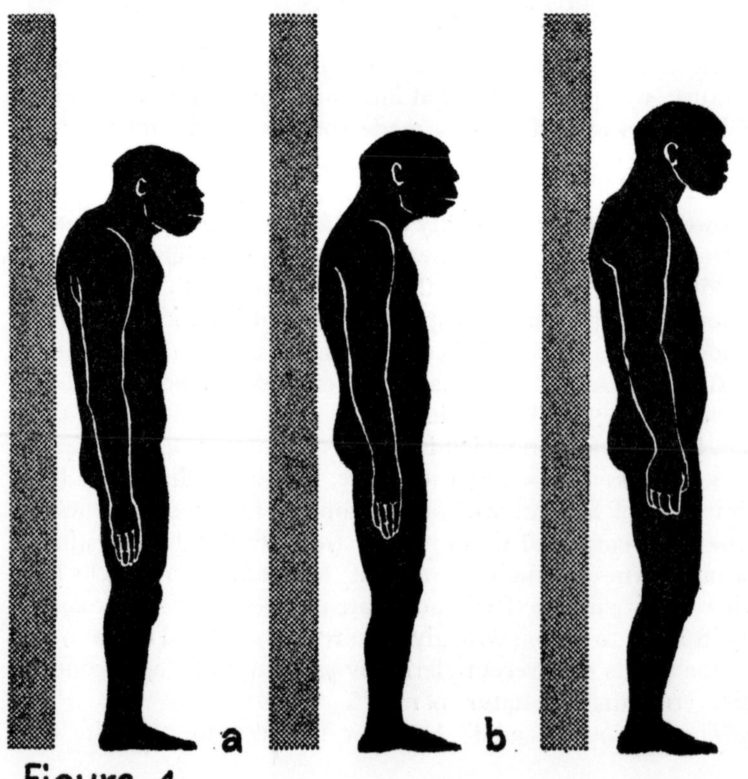

Figure 4

The Upright Balance

The earliest man-like creatures had a short neck and a well developed "hump". Figure 4 shows diagrammatically an evolution from (a) Proconsul man 2 million years ago, to (b) Pekin man 500,000 years ago, to (c) Neanderthal man 100,000 years ago, to (d) Mount Carmel man 40,000 years ago via (e) Modern man to what I might perhaps call (f) Alexander man! One of the most striking features is the way in which the neck is gradually lengthened and the "hump" has become less prominent. In the process, the centre of gravity has shifted backwards, and the point of skull-balance (the occipital condyles) has come back until the body's centre of gravity now can pass through this point.[8, 9]

As a result, Modern man's neck has become more free to move, but unfortunately a freely moving neck—although giving him a wide ranging ability to turn his gaze and his sensory attention around and about himself—also allows him to

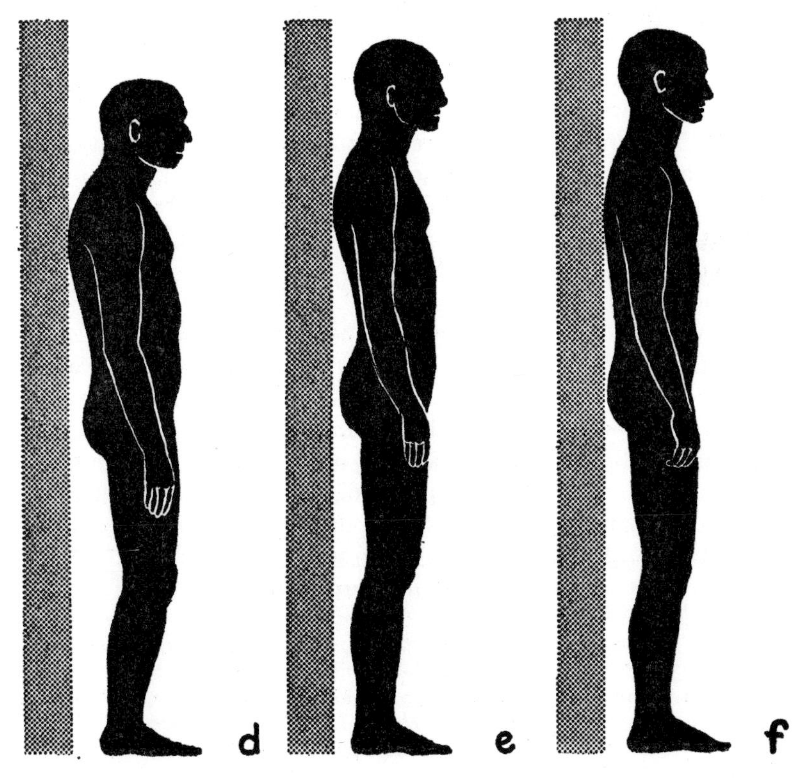

collapse the vertebrae of his neck and his spine. His longer neck allows him more potential freedom, but it also allows him to use certain muscles for activities and postures in which they should play no part. Speaking, swallowing and even ordinary breathing are often made to involve muscle groups in the neck and hump which contribute nothing to the act being performed. The more flexible balance which should be possible is often not actualised: indeed, the potentially free muscles are held fixed in a mis-used balance.*

Alexander balance is selected with a view to promoting the most efficient functioning. Just why it is thought to promote the most efficient functioning will be considered later. At present let us notice in what respect it differs from the balance which most people adopt in the standing position.

The first thing to notice in Alexander balance (fig. 4f) is that, compared with Modern man (fig. 4e), the whole vertebral column is carried much further back. A plumb line from the mastoid process falls through the trochanter of the thigh bone and slightly behind the malleolus of the ankle. But not only this: instead of the neck vertebrae and lumbar vertebrae dropping forward and downward, they are directed up-and-back. Not to the point where they are overstrained, but to a point at which excessive muscle tension in the neck and lower back is released. The effect of this is to increase a person's height slightly in younger people, and considerably in older people, who often have collapsed two inches from their younger height by the time they are 50.

It will also be noticed that the knees are held slightly flexed and the pelvis released so that the pubis points more towards the front. The sexual organs, instead of being pointed to the floor, with associated buttock tension (the "frozen pelvis"), are presented slightly more forward—not by swaying the pelvis forward, but by tipping the pelvis slightly on the lumbar spine.

In this balance, the surfaces of the vertebral joints tend to separate, rather than to be contracted towards each other. Indeed, the Alexander balance throughout the whole of the

* See W. Le Gros Clark, *The Antecedents of Man* (Edinburgh University Press, 1959), in which he calculates the position of the condyles geometrically, giving a rating of 30 for Pekin man, 40 for Australopithecus, and 80 for Modern Man. Also B. Campbell, *Human Evolution* (University of Chicago Press, 1967), in which he shows the same process in primates.

body—shoulder blades, shoulders, elbows and hands, hips, knees, ankles and feet—seeks to establish a resting position in which all the joint surfaces are lengthening away from each other. And since it is more and more realised now by neurophysiologists that awareness of muscle balance derives from an awareness of the *lengthening* of muscle, a newly-ordered "Wisdom of the Body" is likely to be facilitated by such lengthening.

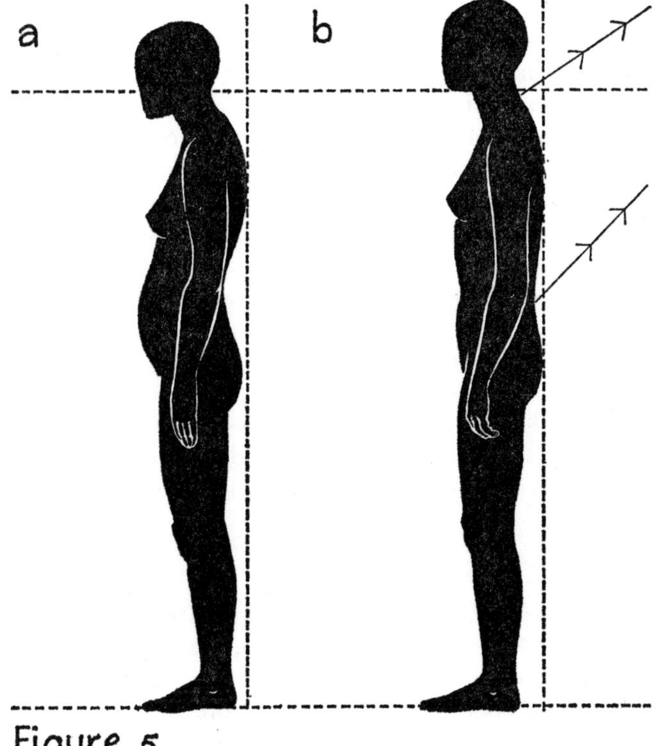

Figure 5

Plate 12 shows a patient who suffered from tension headaches: it shows him before and after he had learned to apply the Alexander Principle. 12a and 12b show him in profile, 12c and 12d show him photographed on the same occasion from the back view. The most obvious change is that he is taller: and he is wider across the shoulders. Contracted shortened muscles have lengthened.

In 12a his weight is thrown forward: his neck is dropped forward and his back is arched. In 12b his back is better although still too arched. Observe the small triangle made by the lines of the grid in front of his throat, and observe how the triangle grows as his neck comes back.

Plate 12c shows him from the back. Observe the lines of muscle contraction at the back of his neck: the raised tensed shoulders and the tightened buttocks with exaggerated dimpling. Plate 12d shows him when he has learned to lengthen the contracted neck muscles and to widen the shoulders apart. His buttocks are now uncontracted and he is appreciably taller. And free from tension headaches.

Figure 5 makes these points diagrammatically. The neck in 5a is dropped down and forward, the back is arched, the pelvis is tipped so that the abdominal contents fall forward. In 5b, the direction of the lengthening has been sketched in. The lines are *up* and *back*, both in the neck and in the lower spine. The head is not pulled backwards and down into the chest: the shoulders are not hunched.

Plate 10c shows these faults in a middle-aged man. The neck is dropped forward. The back arches, the pelvis is tipped forward. The French word for the pelvis—*bassin*—should remind us that the pelvis is shaped to contain the abdominal contents, not to let them slop forward over the front of the *bassin*.

Sitting Balance

The same principles apply to the sitting position. It was noticed in the last chapter that most people, when they sit down, contract the head into the shoulders and as they descend usually arch the back and thrust their bottoms out. There is of course the alternative method of hurling the body precipitately into an easy chair, with the back flexed into a round ball so that the buttocks land on the front of the seat and the backbone curves along the rest of the seat and up the back of it.

But notice what happens when you sit down slowly. What should happen is that—with the heels apart from each other and toes turned out—the knee cap should move continuously forward over the line of the foot (pointing approximately between the big toe and the second toe). As the knees move forward the body will begin to descend.

At this point (plates 1 and 2a) most people (fig. 6a) will:

(i) pull their heads back
(ii) throw the lower chest forward
(iii) throw the pelvis backwards.

Instead the body should descend between two vertical lines (as in fig. 6b). The pelvis should not push back and the lower chest should not push forward.*

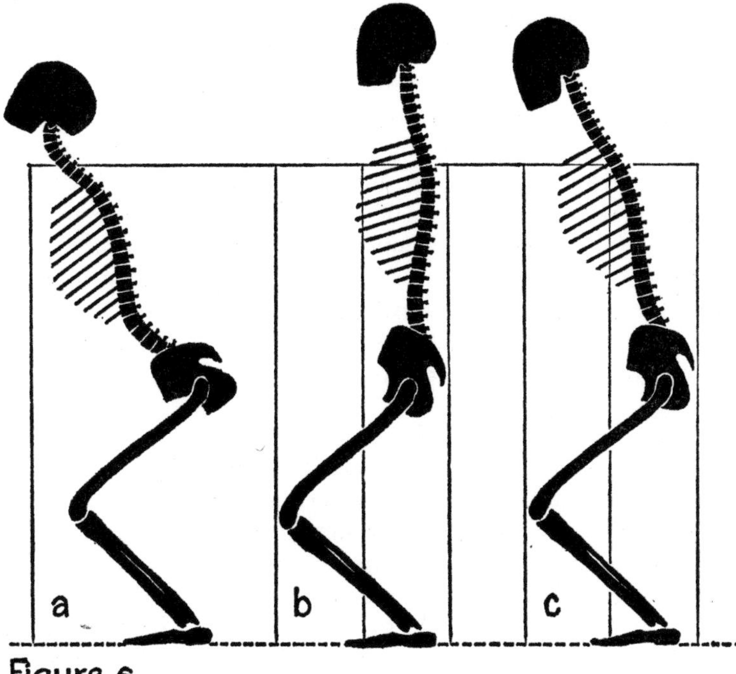

Figure 6

Depending on the height of the chair, the vertical axis of the body can then move backwards in space. Most people at this point fear that they will lose their balance and fall backwards if they continue on downwards. This, however, will not happen, provided that the head is not allowed to tighten back. Instead it must be directed forward at the top of the neck. When the

* This whole movement can perhaps be better understood by reference to a similar movement of the back when placed against a wall, as described in Chapter 9, figure 25.

USE is very wrong, it may be necessary (fig. 6c) to bend slightly forward at the hip-joint. This movement is shown in plate 13a.

Now for most people this is not at first easy, except perhaps in drama and movement colleges, where they have to think about such body mechanics in detail. An amusing note from one of my patients puts most of the difficulties in a nut-shell. He wrote:

> There seem to be two kinds of sitting down—falling into a chair and the kind that can be halted. You point out that in the second kind, there is usually an alteration in the position of the head so that it is thrown back. The thing I don't understand is why there should be an implied disapproval of this backward head motion. The kind of sitting down that can be halted involves maintaining one's balance until the arse is firmly on the seat, and *that*, in my misguided view, entails keeping the centre of gravity vertically above the feet. If one follows your instructions, keeping the head and back vertical and bending the knees, the pelvis comes downwards on the heels and would miss any chair behind one.
>
> It seems to me that in sitting down one combines three actions:
>
> 1. bending the knees to lower the pelvis
> 2. sticking the arse out backwards to get it above the chair.
> 3. bending the body forwards to counter 2.
>
> Now if in 3 the head is brought back in relation to the body, several advantages are secured: one's line of sight, normally horizontal, is so much less disturbed than it would be by pointing the nose to the ground, and one's face is maintained at the normal angle of view to one's companions.
>
> In short it seems to me that there are the soundest reasons for pulling the head back on sitting down in the controlled, as distinct from falling, manner. However, on seeing you on your feet or in a chair I'm completely convinced that you have "got something", without a doubt. But I can't see any reason for suggesting that we should not pull our head back when sitting down.

I was able to show this patient that, if he constantly drew his head back into his shoulders when he sat down, then he would arrive in a wrong seated position in which the head was held hunched into the shoulders as in plates 4 and 28 and that this would lead to "hump" formation and the associated muscle tension and mis-use which this implies: and that, by moving in this way, the spine gradually comes to be shortened, rather in the way which a string of beads (fig. 7a) is straight when lengthened, but goes into curves when it is shortened (fig. 7b). I was also able to show that in fact he could plant his "arse" perfectly well in the manner which I described, provided he kept his knees moving forward and let the vertical axis of his body displace back. And that it was then a simple matter, if necessary, to sit further back in the seat.

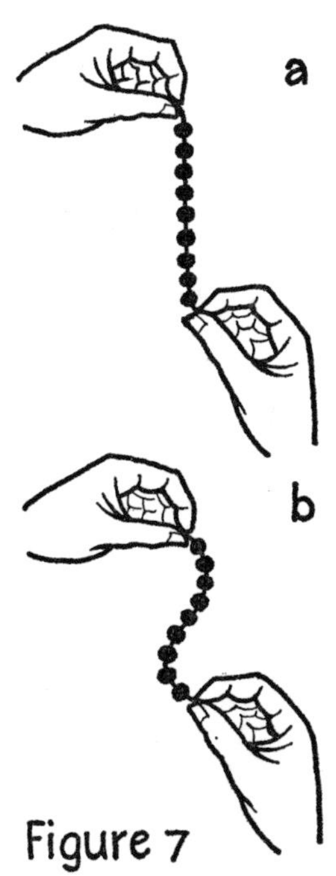

Figure 7

Head Balance

It should not be thought that the Alexander head balance is simply a matter of idiosyncratic choice. There can be few anatomists and physiologists who do not now accept—in theory at least—the importance of head-position, but the lay reader may be interested in the vestibular apparatus in the inner ear which gives us much information about our balance and about variations in pressures which act on our bodies.

The vestibular apparatus (fig. 8) lies inside the skull, internal to the mastoid process, and it registers variations in pressure both from the outside and from inside the body. When we stop or start moving, when we lean on things, or when we fix one part of our body closer to another part, this apparatus should help to tell us what is happening. Likewise it gives us information about our spatial orientation and about the way we are supporting our body against gravity on various surfaces—our feet on the ground, our buttocks on a seat, our back when lying down.

It does this by means of built-in spirit-levels—the so-called "labyrinth". The names of the particular parts do not matter; the essence of it is that there are cavities placed at right angles in three planes in the skull, which are filled with a heavy gelatinous fluid, and in contact with this fluid there are a number of hairs projecting from the cavity walls.

The weight of the fluid drags on the hairs in accordance with the head position, and, as we move or rotate our bodies, the inertia of the fluid jogs the fluid up and down against the hairs. The cavities also have a small flap—a "cupula"—which sways like a swing door to and from a resting position. All of this sends information to the brain about body positionings in the vertical and longitudinal axes—up/down, right/left, front/back—and it also gives information about acceleration and deceleration, by displacement of the gelatinous fluid. It does this more accurately if it is carried on a symmetrically balanced head.

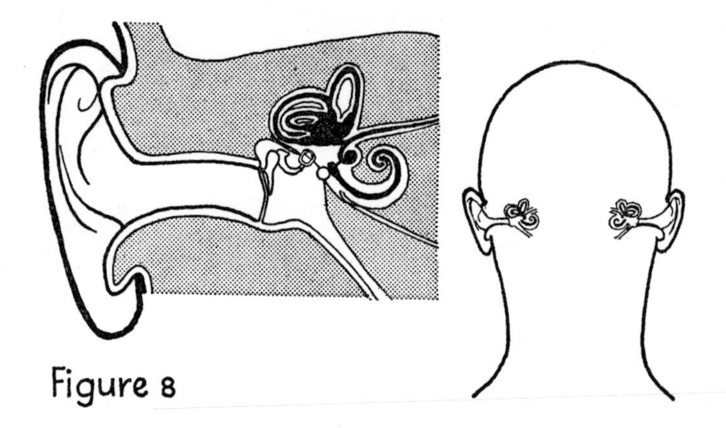

Figure 8

As well as the vestibular apparatus, the skull of course carries the eyes, which also give a sense of position and acceleration. But in modern civilisation, the eyes are frequently being dropped to read or to write or to perform manual tasks. This dropping of the eyes very soon involves a habit of dropping the head forward from the hump, and this position may be held for long periods. As a result, when the eyes are raised (see fig. 22c, Chapter 9), the tendency is for the head to pull back at the point where the neck joins the skull: and thus the hump is perpetuated and the head is further encouraged to be held retracted at the top of the neck.

A correct resting head balance, in which the vestibular apparatus can be carried on an even keel, provides a stable platform from which the special senses—eyes, mouth, nose and ears—can all work. All too often, the vestibular apparatus sacrifices its primary position to the demands of the other senses—the eyes to focus on (or reject) certain sights, the ears to pick up (or to block out) certain sounds, etc. Our precariously evolved head balance is easily disturbed by the bombardments of the modern world and by our incessant desire to pick up or reject information through our special senses. It becomes a prime necessity that we should re-establish a balanced resting position for the head.

Slump

Alexander-USE suggests then that, if adequate functioning is to be maintained, the head balance should not be upset and it follows from this that the back should not be allowed to slump (plate 14a) when sitting. This for civilised man is a tall order—at first many people find it hard to maintain a lengthening balance when seated. Lengthening is often wrongly interpreted as the need to sit up (plate 14b) over-straight, with the back arched and the chest pushed out and with the weight carried through the upper thigh instead of through the ischial tuberosities—the two small knuckles of bone at the back of the pelvis. Yet, in fact, a correct lengthening balance (plate 14c) is restful and efficient, and soon comes to feel comfortable, once the habit has been acquired.

Not only when sitting up straight but when leaning forward to eat or to write or to read, the back should not be collapsed forward from the hump but should be lengthened and then pivoted forward from the hip joints so that the pelvis moves with the rest of the back. In this way, there will be no excessive slumping in the lower back and the whole trunk will not collapse forward.

In addition to the lengthened use of the trunk, the knees should never be crossed when sitting. Whenever it is socially possible and there is enough room, the knees should be pointed away from each other. Most forms of lower back pain will be benefited by directing the knees away from each other, and this particularly applies to sedentary workers, who sit at desks all day long.

Once you have sat down, it is usually best to move your pelvis deep into the back of the chair seat: this applies to almost all seats: cinema seats, buses, trains, dining chairs and easy chairs whenever the leg-length allows it. Unfortunately many modern chairs—particularly those with a marked curve between the upright and the horizontal—make good USE almost impossible.

The television habit has led to a great deterioration in children's sitting posture. Television, whether we like it or not, is with us to stay, in one form or another It should not be too difficult for parents to encourage their children not to collapse and slump as they watch the screen. If children are too tired for this—after school for example—it is best to arrange for them to lie down with their backs supported, rather than to sit slumping.

Children, when shown, are able to maintain the balance I have been describing without strain and effort, and they will become less tired by their school work as a result. The forces in their environment do however conspire to teach them to mis-use their bodies: the nature of these forces will be elaborated in the chapter on Personal Growth.

Alexander-balance may at first sound as if it entails an immense effort. The learning of it when it has been lost will certainly entail effort; but, when learned, most people find it feels so good and easy that they would not willingly throw it away.

Balanced Wisdom

The up-shot of the matter is that not only are we poorly designed for achieving balance—poor, in the sense that it does not come easily and automatically to us—but also we have had little instruction in how to get the best out of our particular model. Secretaries and car-drivers have had plenty of scientific investigation into the design of their desks and chairs and their car seats: but in common with all of us, they have been expected to know how to make the best they can out of the body-work which they got from Mum and Dad. And even though they may cast envious eyes at the muscular and bustular torsos which fill their screens and newspapers, they have not learnt to feel an envious thrill at the sight of a properly poised pelvis or a sensibly seated sacrum. They may appreciate the fact that a Dressage rider needs to sit well so that he can control his horse by his buttock-adjustment: but a sensitively poised sitting position has not seemed important to them in the idle boredom of their

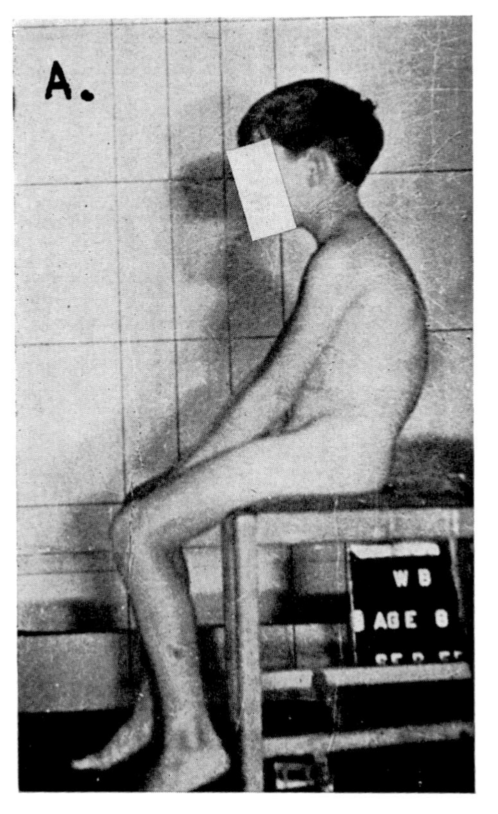

14a Slumping.

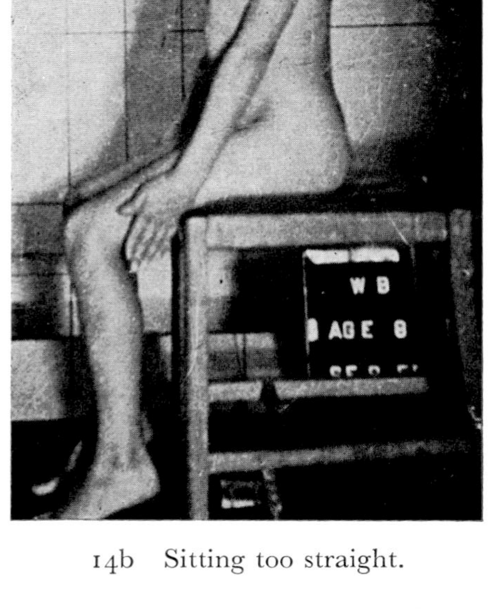

14b Sitting too straight.

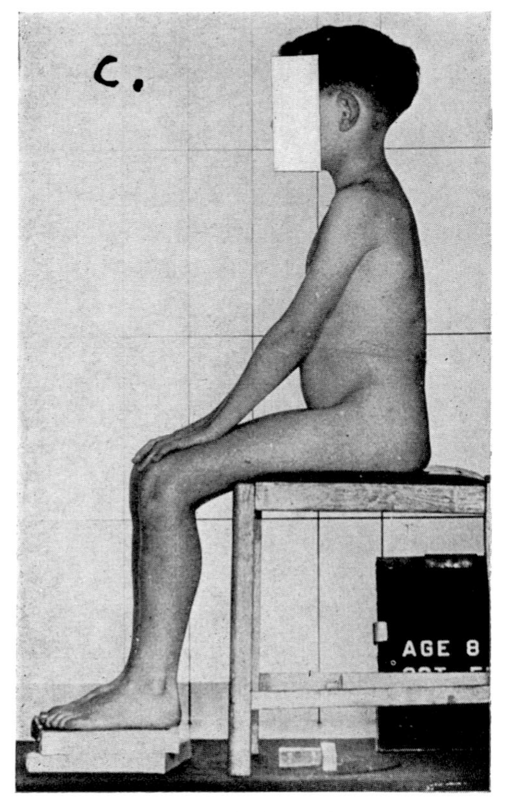

14c Balanced.

15 Muscle-spindle lengthwise (above). Cut through vertically (below) to show small internal muscles and large external muscles.

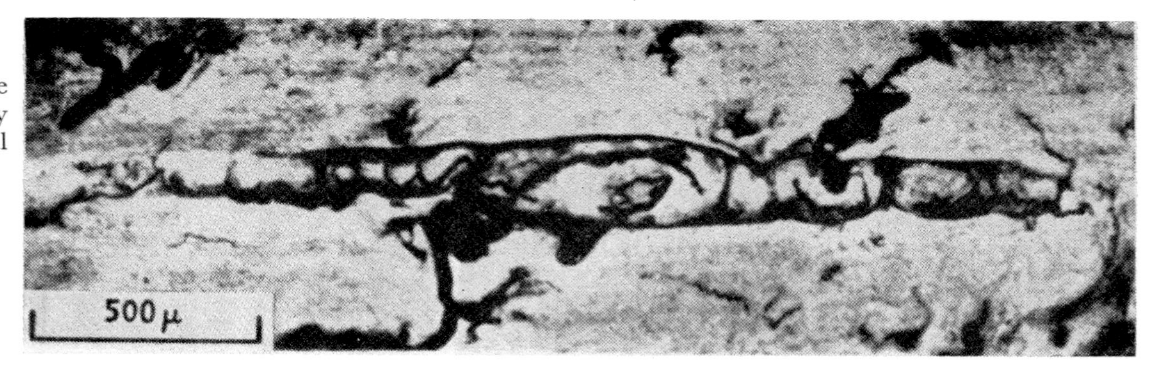

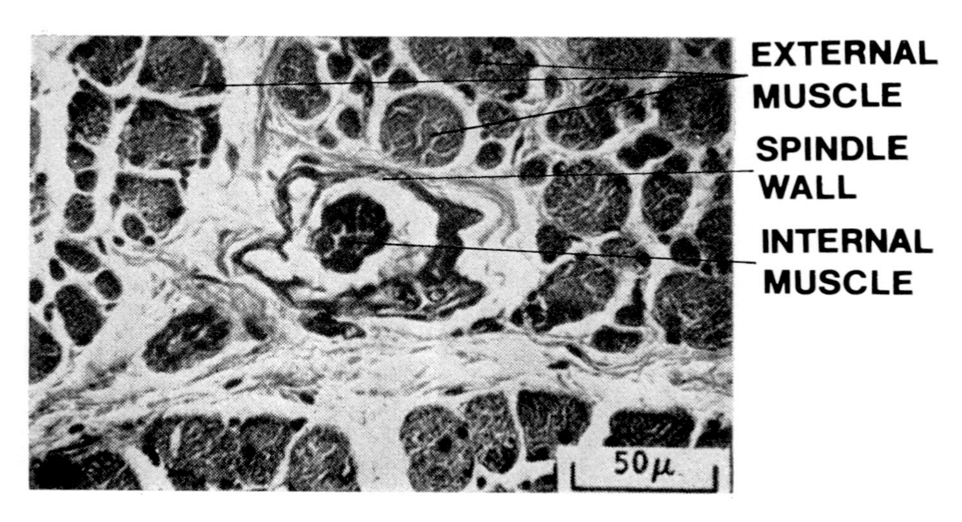

office, or during the stupefying collapse of politicians on their Parliamentary benches and in their committees. And even the most famous of athletes do not seem, when sitting, immune from a state of collapse which ages them prematurely and eventually nullifies their potentialities just when their skill and experience should be bearing further fruit.

It may sound as though Alexander-USE is simply an elaboration of all the good advice on posture which has been poured out for years in physical education manuals, health and beauty movements, and, more recently, in westernised meditation regimes. But what I have written is only the beginning of a study which involves our most personal and most intimate adjustments in every day living—a study which begins to make sense of phrases like "the whole man" and "psycho-somatic integration".

The further implications of Alexander-USE involve a consideration of the meaning of REST, and to this we must turn in the next chapter.

Chapter 4

REST

ALL OF US, in the past few decades, have been through a period which has been characterised by immense change—a change in our means of living as well as a change in our modes of living. During this period, the *tempo* of change, already quickening in the first half of the century, has accelerated at a pace quite out of proportion to the pace of man's earlier evolution. Human relations in almost every sphere have become characterised by speed, by too many things being done far too quickly, without sufficient time being possible for the necessary evolution of biological sequences. No one is immune. Frank Sinatra recently put it well when he spoke of "The need which every thinking man has for a fallow period in which to seek a better understanding of the vast transforming changes now taking place everywhere in the world."

Under these circumstances—to which the customary phrase "the stress of civilisation" scarcely does justice—a consideration of the meaning of REST has become of vital importance. Indeed, the diagnosis of the problem which confronts civilised man is not new. It is, essentially, the need of every man to find biological harmony in a world far removed from conditions suitable for adequate biological functioning. This is a problem which we all of us face—how to live in a confusing and quickly changing world without losing our biological harmony, and without losing satisfaction in our daily living. Such "biological harmony" is impossible without the ability to achieve a balanced state of rest, as opposed to the state of "dis-ease" and fatigue which for most people follows a stress-activity.

There has been a tendency recently to speak as if the main battle against disease and ill health has already been won, and to suggest that an extension of our present methods of prevention and cure would solve the problems of health. It is easy to point with pride to the decline in infant and maternal mortality rates, to the mastery of most of the infections and to our greatly increased expectation of life as compared with a hundred years ago. But it is unduly naive to believe that increases in height or weight or in length of life, are proof that the day-to-day health is satisfactory. Certainly the dramatic forms of disease, such as cholera, smallpox, and typhoid, have been almost eliminated in Western civilisation. But we are left with the less dramatic states of "dis-ease", of departures from normal health and from a balanced state of rest. The medical profession has scarcely begun to explain what it is that distinguishes an individual in full health from an individual with "no demonstrable disease".

Health involves many things at many levels but full health is impossible unless we can maintain a balanced equilibrium in the face of forces which tend to disturb us. And this is where the Alexander Principle comes in. Alexander drew attention for the first time to the structural conditions in which a balanced equilibrium is possible, as opposed to those in which it is bound to be impossible. He also showed how most of us are encouraged and even taught to move and react in an unbalanced manner, until eventually we reach a state of affairs where we not only cannot recognise how we are mis-using ourselves, but do not even *want* a different USE when it conflicts with our habitual social attitudes. A state of affairs where we do not know how to achieve a balanced state of rest.

Posture

Unfortunately, a markedly wrong imbalance may not affect biological functioning in an obvious way at first, and this accounts for the bizarre variations in posture which come to be accepted as normal and suitable for particular social situations and surroundings.

There is no one single social criterion of what is a good posture—it has many different meanings to different people. The barrack square sergeant, the nanny, the anthropologist, the dancer, the gynaecologist, the sculptor, the actor, the Buddhist monk—these and many more will all have their particular ideas of what is right: the adolescent thinks it essential to adopt the typical slouch: the model shows off her clothes with grotesquely thrown forward pelvis: the shop assistant and the bar-drinker relax with weight on one leg: the professional beauty queen arches her back and pushes out her bosom: and so on through the whole range of body-language which we may think suitable and appropriate. None of these postures would matter too much if their perpetrators had some idea of a postural norm to which they could return when the immediate pressure of the social moment was over. But these distortions *become* each person's own norm, and feel so right that a properly balanced use of the body may feel unnatural. Momentary attitudes in time become habitual dispositions and the body soon becomes moulded into fixed patterns which to a large extent will determine future performance and future functioning.

It is clear that we cannot rely on a *social* criterion of what constitutes good USE. *It is from their effect on biological functioning that the variety of body uses must be judged.* There are many alternative possibilities in the mechanical use of the body at any given time, but there is, for any given situation, a way of using the body which makes for the best functioning, for the least wear and tear, and for the sweetest running engine, just as there is a USE which leads to waste of energy and undue fatigue.

Many writers before Alexander have written of this need for ease and economy of effort. Schopenhauer[10] considered that the USE was good if every movement was performed and every position assumed in the easiest, most appropriate and convenient way—"the pure adequate expression of intention without any superfluity which might exhibit itself in aimless meaningless bustle". Unfortunately the "easiest and most convenient way", although perhaps socially appropriate and convenient, is not necessarily biologically appropriate. Herbert Spencer[11] perhaps came nearer to it when he spoke of "movements which are effected with economy of force, and postures which are maintained within this economy". Likewise Marcus Aurelius wrote: "The body ought to be stable and free from all irregularity whether in rest or in motion. All this should be without any element of affectation." Thomas Aquinas in *Summa Theologica* thought that good use consisted in "due proportion, for the sense delights in things duly proportioned: delight springs from evidence of ease in the performer".

Yes indeed, but what is Easeful USE? When mis-use patterns are never relinquished, but are present even at rest, we are confronted with a state of pervading dis-ease and strain which stops life from being lived as it should be. The fact is that the majority of people do not really know how to achieve an easeful state of muscular rest in their bodies. When their childhood remedy of a good night's sleep fails to restore them to full enjoyable energy, they seek for artificial sleep and artificial tension-release: and the drugs which they use to relax their muscle-tension produce a state of dullness which is a mockery of what living should be.

Dystonia

The medical name for faulty muscular tension patterns is "dystonia", and it will simplify description if I refer to such patterns of muscular mis-use as "dystonic" patterns.

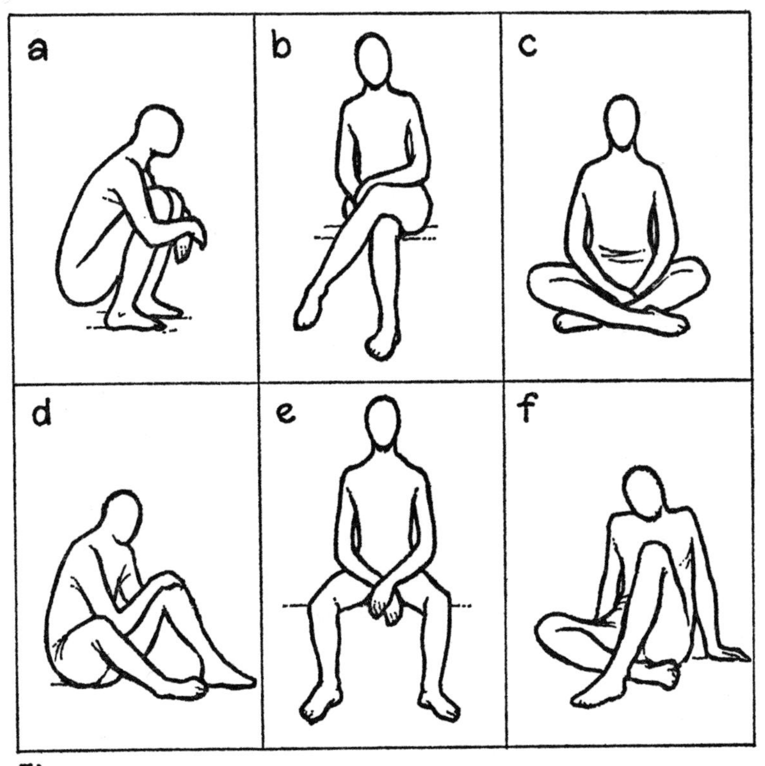

Figure 9

Dystonic patterns arise and produce an unbalanced resting-state in many ways. They are particularly obvious in the positionings and postures which we adopt when we are keeping still. Over one thousand such body positions have been listed, all of them variations of sitting and lying and standing and kneeling. Some of them seem unusual to western eyes, but the deep squat (fig. 9a) and the tailor squat (fig. 9c) employ a far better USE of the back than, say, the familiar adolescent postures of figure 9d and figure 9f. Figure 9e in which the legs are not crossed is functionally far better than the familiar crossed-knee position of figure 9b.

Dystonic patterns also arise in the simple mechanical actions which we carry out all day and every day when we move ourselves and move objects in our surroundings: moving a fork, a book, a paper, a telephone: moving a dish cloth, an oven door, an electric switch: moving a gear handle, a coin, a bus ticket . . . the list is endless. Dystonic mis-uses appear when we walk and run: when we jump, hurdle, swim, throw, dance; when we swing a golf club, a tennis racket, a cricket bat, a conductor's baton: when we ride on a horse, or a bus, or a bicycle: when we sit slumped in a dinghy or huddled over a text book in the library: when we lift a dish or a dictionary from a shelf: when we stand at a bar or in a shop or at a football match: when we carry out surgical operations or laboratory work or dentistry: when we work manually in industry or agriculture or just in the garden. In all of these, and many more, our performance and our liability to fatigue is bound to be influenced by our manner-of-use.

In activities which need a special skill, muscular dystonia leads to unpredictability of performance. To consider only one athletic sphere, most of our top-class women tennis-players in Britain never quite live-up to their earlier promises. They may have isolated successes, but often, just when their biggest opportunity comes, they become pathetically erratic. Most of them show marked dystonic patterns around the shoulders, upper back and neck, which become accentuated as they come under pressure. As the game proceeds, they can be seen to accentuate their tension, and they do not come back to a proper resting balance.

Of course in the midst of any game—or indeed of any intense activity—there has to be a concentration of effort and expectancy on the matter in hand. But such concentration need not produce dystonia, and we are faced with two questions:

(a) how are we not to mis-use our bodies when we start to do something, and

(b) when we stop, how are we to release the muscular contractions which we have just been making.

Such a muscular control will only become possible if we can start from a properly balanced state of rest, and if we know how to return to (and maintain) such a steady state of muscular rest when we stop. Alexander's concept of USE implied a conscious awareness of such a steady state.

*Postural Homeostasis**

About twenty years ago, I suggested the phrase "postural homeostasis"[12] to describe the steady state in which the body keeps itself balanced. Postural Homeostasis involves a most intricate and delicate interplay of muscular co-ordinations and adjustments throughout the body, to bring the body close to a balanced state.

The balance which results from this inter-play is in what the physicists call "a steady resting state", and in a healthy person these muscular adjustments will mesh together to give a balanced whole: a juggler who balances a number of objects on a pole is maintaining them in a "steady resting state". Work is being done to maintain balance around a central point of stillness. This central point is not fixed. Oscillation takes place around it, with smaller, or bigger swings. Balance can be achieved in all manner of ways—many of them markedly inefficient, with too big an oscillation away from the central resting point.

Such oscillation is characteristic of all our muscular activities. If you look at yourself in a mirror you will see that you are swaying slightly. More obviously if you place a piece of string down the mirror and stand back about ten feet and line up your nose with the piece of string, you will notice as you walk towards the mirror that your nose oscillates a great deal from one side of the string to the other.

Even, when standing still, a pin point of light photographed on the top of the skull (fig. 10) will show a great deal of sway around the central point. Eysenck has confirmed that in neurotic people[13] such swaying oscillations are much larger than in healthy people: indeed, excessive sway is one of the clearest indications of conflict in people's personality.

* This section (from page 51–62), although essential for a full understanding of the Alexander Principle, may be skipped if necessary.

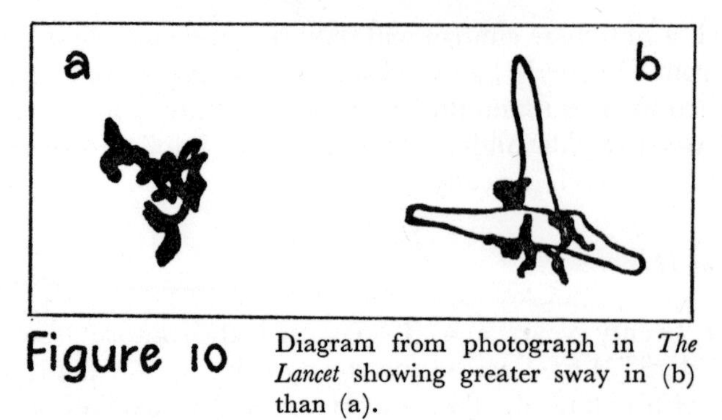

Figure 10 Diagram from photograph in *The Lancet* showing greater sway in (b) than (a).

Muscular Feed-back

What governs the amount of oscillation in our muscular adjustments? Think of a simple movement (fig. 11) like moving the tip of the right index finger (X) to touch the tip of the left index finger (Y). The distance between X and Y has to be assessed by our brains rather as a cat gauges how to jump from a window ledge to a parapet.

The distance between X and Y is known in cybernetic jargon as the "error". Information about the "error" XY is fed back

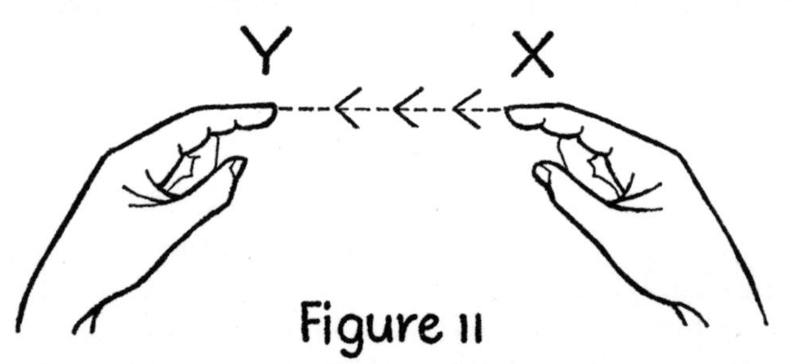

Figure 11

to the brain where it is unconsciously checked against a pre-existing model (the receptor element R, in figure 12). According to the construction which we place upon the information received action takes place in the muscles (the effector element M, in figure 12) to move from X to Y. In other words to close the gap and to eliminate the error.

Figure 12 is the prototype of a homeostatic circuit, and it involves what is known as "negative feed-back". Negative feed-back keeps a system oscillating very close to a central resting state, and when there is too big a movement away from the resting position, it brings it back by a compensatory movement.

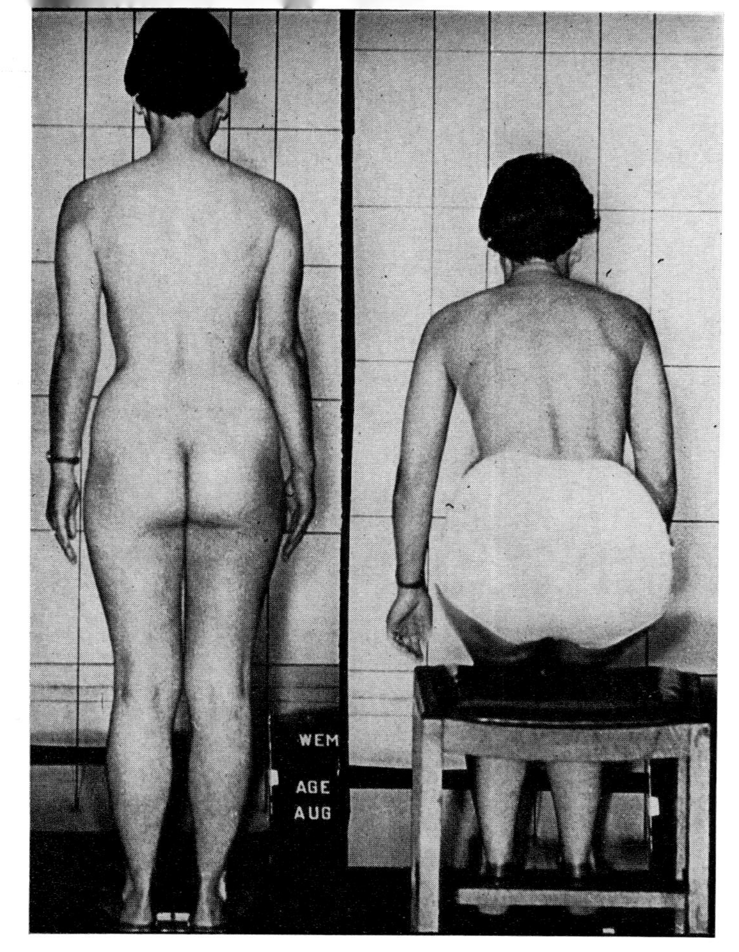

16 Note neck tension at back. Raised right shoulder. Tension left side of lower back. Pelvis twisting to right.

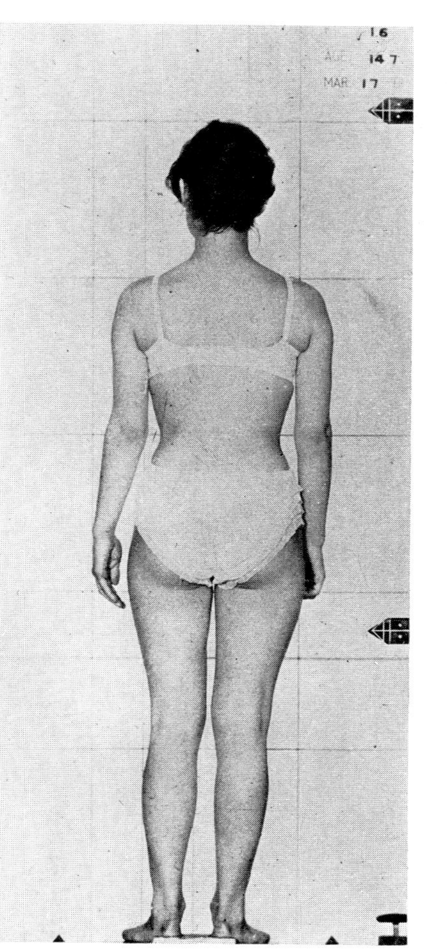

17 Patient standing on right leg. Central upright of grid should pass through middle of back and pelvis.

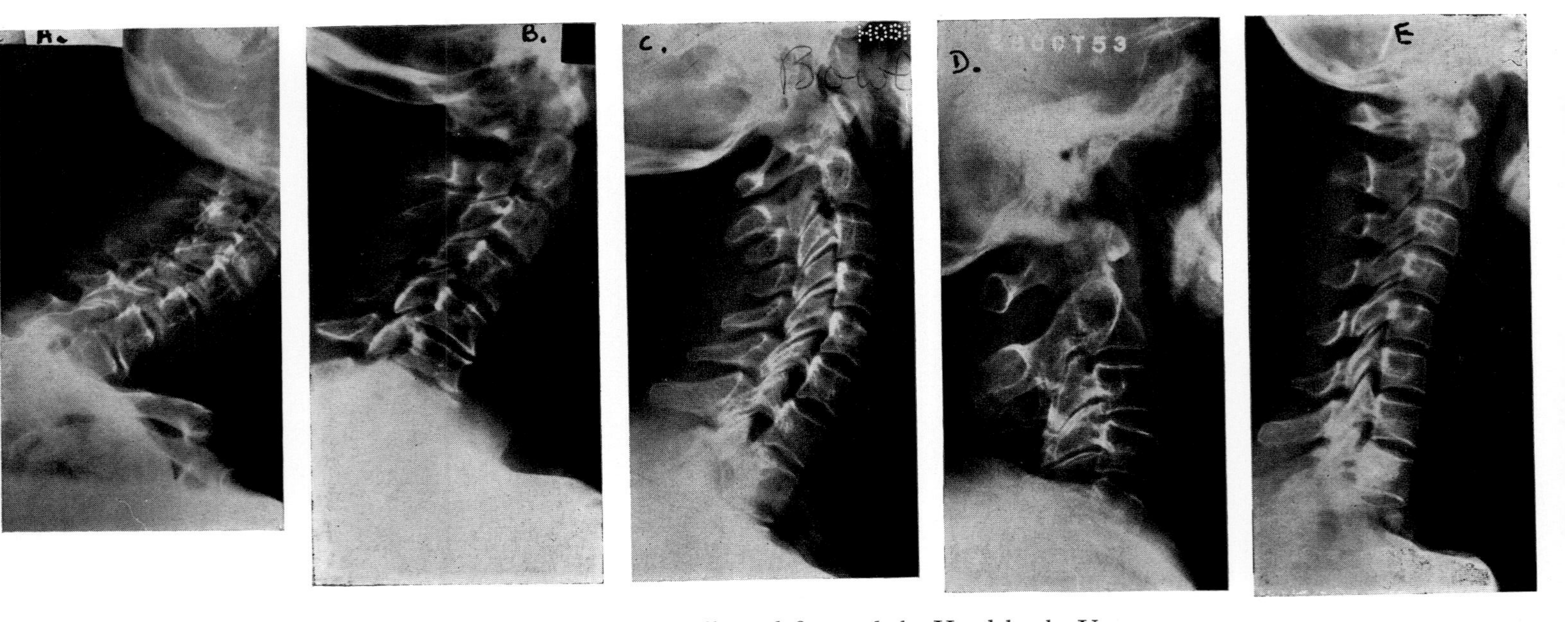

18a, b, c, d, e a. Neck collapsed forward. b. Head back, Upper neck forward, lower neck back. c. Head pulled back. d. Lower neck collapsed out of sight. e. Overstraightened neck. Fourth vertebra slipped forward on fifth.

The receptor element R represents our idea of what—for one reason or another—we consider to be a normal resting state. Through it, we receive information and compute it against our previously stored experience. A useful way of saying this is to say that we all have a "body-construct", made up of all of our previous learned experience of our body, and that we construe

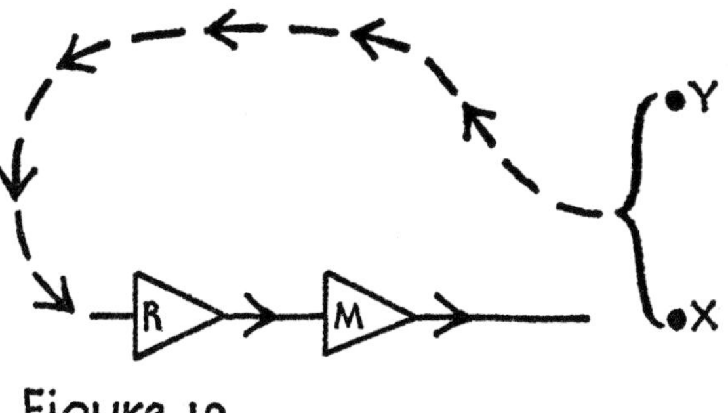

Figure 12

(put-a-construction-on) what happens to us by reference to this "body-construct".

We can also think of "error" as being the difference between the way we construe things to be and the way we want them to be. And of course we all of us may construe things—and alter our perception of them—to make them seem to be the way we want them to be.

The feed-back circuit in figure 12 is similar to the scanning mechanism which is used in television and radar. The muscles are scanned as one might scan a printed page in search of a word. A thermostat is another example: when it corrects an error, it makes a new one: this too it corrects by making a further but smaller error. It is in fact in a state of "steady motion". When the self-correcting mechanism overshoots the mark, oscillation will occur.

Riding a bicycle is another example. When the rider falls slightly to the right, he turns his front wheel to the right which stops his fall but leads to his being thrown to the left, and he corrects this by turning his wheel to the left, and so on. The net result is a steady resting state in his body, as he uses his arms and his legs to steer and to pedal.

If our "use" is to be accurately balanced, four things at least are needed. Firstly, we need to get adequate information from our muscles (and from the other parts concerned with movement). Secondly, we need to receive this information accurately in our brains without obscuring it. Thirdly, we need to activate our muscles so that they do what we want, with a minimum of mis-use. And fourthly, we need to know how to come back to—and maintain—a balanced resting use of our bodies which will interfere least with our functioning.

Muscle Physiology

In a book of this nature, it is totally impossible to give a full account of all that is relevant in muscle physiology. In the time since I took a degree in physiology at Oxford, the complex study of nerve and muscle has become yearly more complex; but except at a fairly crude level of neuromuscular injury or pathology, it is still almost impossible to relate either the old or the new muscle physiology to what actually happens to "normal" human beings at rest and in movement. Perhaps I can cover myself with those who find my account too simple by recommending R. A. Granit's *The Basis of Motor Control*[14] and T. Roberts' *Basic Ideas in Neurophysiology*[15] for more detailed study.

I know of no other sphere of physiology in which acutely intelligent minds have laboured with more imagination and skill. But, in spite of this fine physiological work and speculation, the task of teaching muscular control to actual people in their actual daily affairs has not yet been greatly facilitated. What I have to say about muscle control fits the facts as I know them; and if it is argued that this or that piece of muscle physiology points in another direction, then, rather like Bishop Berkeley, I would reply by saying "I refute you thus", by showing them what actually happens in living human beings in the clinical situation.

Muscle can either shorten or lengthen. It contracts by means of a molecular shortening which pulls on elastic elements in the muscle fibre: and the contraction is produced by nerve impulses, synchronised in numerous relay-stations in the brain and spinal cord. The resultant impulse to a muscle fibre, delivered at a certain intensity to a muscle in a more or less receptive state, makes it contract, i.e., shorten.

But what happens when muscle lengthens? Is it simply that the nerve impulses which made it contract stop firing, so that it returns back to its original resting length? Unfortunately it is not as easy as that.

We have two systems (fig. 13) by which our muscles are controlled by motor nerves (i.e., nerves which go to the muscle from the brain). The first system—and until recently thought to be the only system—works by making muscle-fibres contract and shorten. 55% of motor-nerves look after this activity. The second system, which uses the remaining 45% of the motor nerves, works on a quite different basis. The nerves from this system do not go directly to actual muscle itself—to the actual biceps muscle or thigh muscle which you can touch with your hand—but to a complex structure called a muscle-spindle, lying within the belly of the anatomical muscles. Many thousands of these lie length-ways in the muscles: they are about 8 mm long, bulging in the middle and tapered at the ends. They are concerned more with the lengthening of muscle than with its contraction.

The muscle-spindle has its own set of internal muscles (plate 15) and in addition to the motor nerves which go *to* it from the brain and spinal cord, it has sensory nerves which go back *from* it to the brain and spinal cord. The spindle is a much more sensitive adjuster of muscle than is the actual overlying muscle itself. Its register of length works in parallel with the overlying muscle not only to damp down excessive oscillations during actual activity, but also *to induce a lengthening of contracted muscle after activity*. In the words of P. A. Merton, it "constitutes a follow-up servo, the muscle length tending to follow changes in spindle length".[16]

The whole mechanism is extremely complex: but from the point of view of someone who is attempting to learn a properly balanced use of his body, two major points arise. Firstly, over-contraction and shortening of anatomical muscle may result in the muscle-spindle going silent, i.e., failing to feed-back information to the brain about how much the muscles are contracting. A spindle stops discharging when over-shortening of the main muscle occurs. And secondly, lengthening of anatomical muscles can be brought about not simply by stopping off the activity which originally made that muscle contract, *but by learning voluntarily to lengthen muscles until they achieve a better resting length.*

It would appear that the muscle-spindle plays a very large

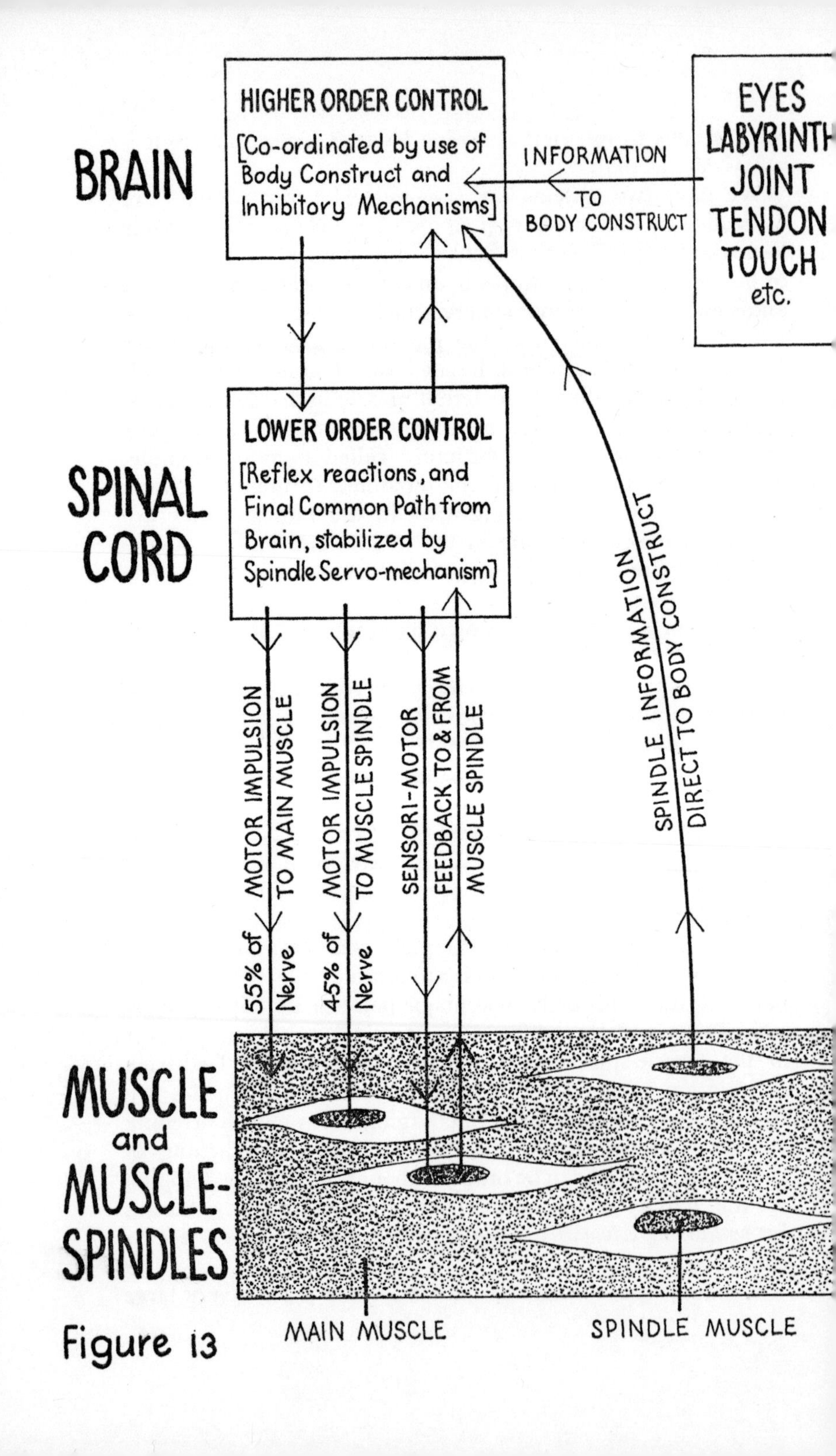
BRAIN
HIGHER ORDER CONTROL
[Co-ordinated by use of Body Construct and Inhibitory Mechanisms]
INFORMATION TO BODY CONSTRUCT
EYES
LABYRINTH
JOINT
TENDON
TOUCH
etc.
SPINAL CORD
LOWER ORDER CONTROL
[Reflex reactions, and Final Common Path from Brain, stabilized by Spindle Servo-mechanism]
SPINDLE INFORMATION DIRECT TO BODY CONSTRUCT
55% of Nerve
MOTOR IMPULSION TO MAIN MUSCLE
45% of Nerve
MOTOR IMPULSION TO MUSCLE SPINDLE
SENSORI-MOTOR FEEDBACK TO & FROM MUSCLE SPINDLE
MUSCLE and MUSCLE-SPINDLES
MAIN MUSCLE
SPINDLE MUSCLE
Figure 13

part in this production of length in contracted muscle: and it should be mentioned that spindles are connected not only with the cerebral cortex (through which we control our actions) but with the reticular formation (the nerve-network in our brain which is responsible for our conscious awareness of the world about us).

We can accordingly learn consciously to lengthen tense muscles, not just by stopping the action which made them contract, but as a definite act of will, by which we can release and re-lengthen contracted muscle.

This, of course, is a much simplified account, and it must be, since any nervous pathway we can trace in the brain ultimately connects up, directly or indirectly, with muscles or muscle-spindles. Any frequently repeated use of a particular nervous pathway is likely to lead to the slow development of a "cell-assembly"—a diffuse structure of cells in the cortex and mid-brain and basal ganglia, capable of acting as an enclosed "memory" system which influences other systems and is influenced by other systems; but this sort of detail is not in place here—the cross-correlations of motor and sensory nerves to and from a vast number of motor units and spindles in a large number of muscles, via the spinal cord and all levels of the brain, is fully described in textbooks. The territory of Cybernetics and Information Theory is already well trodden. But after ploughing through the writings of Wiener, Lashley, Craik, Grey Walter, Von Neumann, McCullough, Shannon *et alia*, my brain still reels at the complexity of it. Marvellous phrases like "Signal/noise discrimination", "Error-minimising neuronal networks" and "Negative contingent variability", jostle round my skull. My own simplified version of how muscular feed-back works seems to me adequate without doing damage to the facts. Broadbent,[17] in his introduction to the recent *British Medical Bulletin on Cognitive Psychology* (Sept., 1971), reckons that Psychology has only now reached the point when "it can usefully tackle many problems nearer to ordinary human concerns". No doubt the Alexander Principle will before long be greatly clarified and refined by new paradigms of associative mechanisms in perception: but meanwhile the practical business of detecting and re-educating faulty use-patterns can proceed quite satisfactorily with the simplified view which I am suggesting.

Faulty Resting Balance

The patient in plate 16 shows a dystonic pattern both when she is standing upright and when she sits down. The slight twist of her pelvis to the right when she stands becomes much greater when she sits down: and it can be seen that when she sits down she makes an excessive muscle contraction on the left side of her

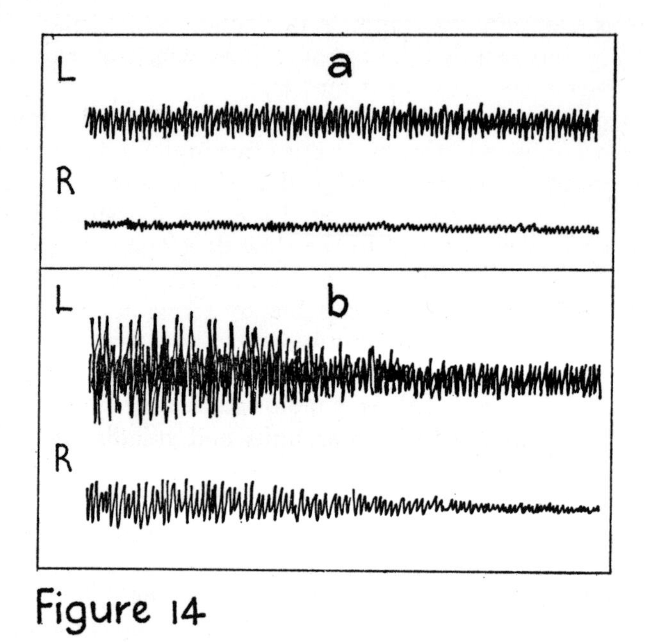

Figure 14

back (just above her pants) which is an accentuation of a similar contraction when she was standing.

Such muscle patterns can be recorded electrically. Figure 14 is a diagram from a recording which I published some years ago in *The Lancet* (2, 659, 1955). In A, the left side of the back shows considerably more activity than the right. In B, when she is moving, there is activity on both sides of the back but more in the left. Her back, in fact, shows a dystonic pattern which distorts the posture of the back, and, in her case, produced pressure on one of her lumbar discs. But, in addition, her resting position was an unbalanced position. Her balance around a central resting point showed excessive oscillations, because of the asymmetry in her muscular resting position.

Figure 15 is a recording of the neck muscles of a violinist who consulted me because of painful cramps at the base of his neck when he was playing fast passages.* In figure 15a the muscle is relatively quiet; in figure 15b he picks up his bow and puts it down. In figure 15c he again picks it up and puts it down, but

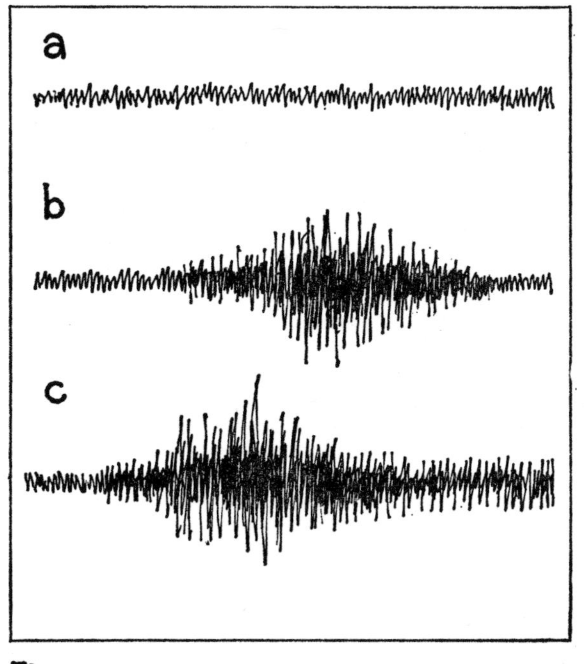

Figure 15

this time the muscle continues to be in a state of activity even when he is doing nothing. His feed-back mechanism has gone wrong, and, at this point, he has no idea of how to release the contraction, but has to rely on it eventually subsiding—an impossibility while he is giving a concert. It is clear that he needs to be taught how to return at will to a properly balanced resting state.

As a simple illustration of the resting state principle—a patella hammer (fig. 16a) which has a heavy head on a flexible wooden handle, will, if agitated, oscillate around a resting point (fig. 16b): application of an external force (fig. 16c) will

* This diagram is taken from a recording in "Posture and the Resting State".[18]

deform it, and, after removal of this force, it will either return to its original resting state or it will remain to a greater or less degree deformed (fig. 16d). In time, repeated application of such a force will lead to structural alteration or, at any rate, to a predisposition to bend more easily, just as a piece of paper, once folded, tends to bend more easily in that direction.

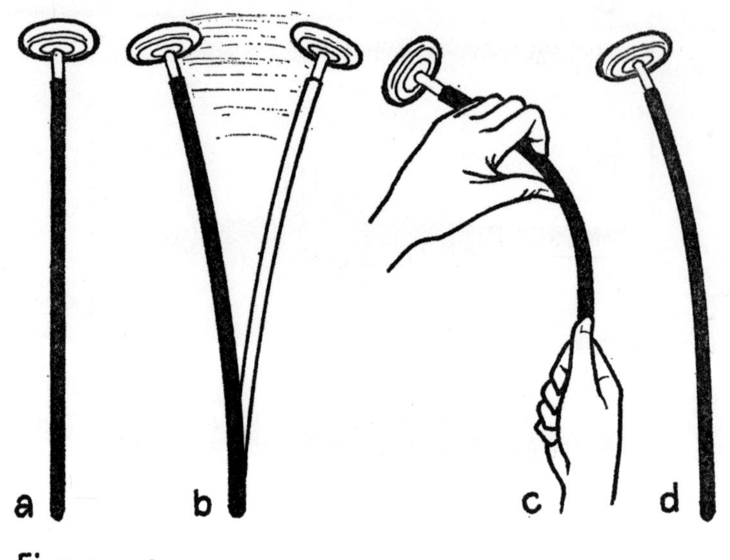

Figure 16

Dystonic mis-use is present when we do not know how to return to a balanced resting state after reacting to a given situation. Such over-active states in time become habitual and a predisposition to adopt them will persist even when they temporarily disappear.

In time, not only does the resting state of the muscle become wrongly balanced, but it begins to modify the bones and joints on which it works and also the circulatory system which traverses it. The bony framework becomes warped and cramped and stretched by the stresses and strains which are put on it by persistent over-contraction of muscle. These states of over-contraction, large and small, gradually leave their mark on us until our resting state is in its various ways as deformed as the patella hammer in fig. 16.

Relaxation

There have been many recipes for getting rid of such unwelcome tension: and most people can get by for quite long periods by one means or another, even though they begin to show persistent dystonia. Most people, by the time their adolescence is over, if not before, begin to feel that something is amiss, and they may already be showing mental or physical disturbances which either produce symptoms or impair their performance.

As we grow older, the resting state which we all of us learn to adopt is either a balanced or an unbalanced one, according to the degree of mal-distributed muscle tension. Muscular hypertension is the residual tension and postural deformity which remains after stress activity—or after any activity which leaves behind residual muscle tension. Such residual tension should, ideally, be resolved by returning to a balanced resting state; but usually it is only partially relaxed, without the dystonic pattern being resolved. In the latter case the tension remains latent in an unbalanced resting state, so that it may only require the idea of moving to re-activate the muscular hypertension, usually in the form of anticipatory tension or "set".

Two alternatives therefore are possible. Firstly, the muscle tension can be relaxed a little without resolving the underlying dystonic pattern, which remains latent until the conditions for its renewal occur. Secondly—and this is what is needed—the tension can be resolved by a return to a balanced resting state, leaving no unconscious residue.

Owing to a lack of understanding of what is entailed by a balanced resting state, most people, with or without the help of their doctor, resort to the first procedure. Under such titles as relaxation, rest-cure, tranquillising or simply by such expedients as alcohol, nicotine and slump-collapse at weekends, a temporary diminution of disagreeable tension may be obtained. And, by avoiding a stress situation or the memory of it, a sufficient number of "funk-holes" may be discovered by which activation of latent tension is avoided. The resultant lack of vitality, creativity, and enjoyment of forward planning (which would activate the tension state) eventually becomes the normal way such people spend their lives.

Why is it that the balanced resting state should be so difficult to attain? Firstly, there is widespread ignorance of what it entails. Secondly—and this will be the topic of later chapters—when someone is in a tension state, there is fear of return to a state of rest. Our preferred self-picture is usually sustained by USING ourselves in a habitual way. By maintaining an unbalanced USE we can maintain self-deception.

The relationship of USE to Mental Health will be discussed in greater detail in Chapter 7, but the significance of Alexander's discovery must first be seen in the sphere of General Medicine. To understand the Medical niche into which his discoveries fit, the concept of Diagnosis itself must be considered: against this background, it will then be possible to detail some of the physical disorders in which USE is particularly relevant.

Chapter 5

MEDICAL DIAGNOSIS

THE WHOLE WORLD of medicine and disease is fraught with fear. The television revolution may have taught us that doctors are fallible human beings, but in moments of dire need, the sheer fear of illness and pain invests one's doctor with superhuman faculties and wisdom. The patient who, say, is unable to pass water and whose full bladder is a screaming agony, would give at that moment, much of his personal fortune to someone who can relieve him. Most doctors will have experienced from such patients a depth of gratitude which affirms them in a view of themselves as basically useful people.

The doctor, with his *therapeutic* skill, will always be loved, feared and respected. But such rewarding therapeutic moments do not happen very often. A vast amount of general medical practice is concerned with humdrum palliative measures, with relatively dissatisfied and grumbling sick people, and with a mixture of kindness and counselling, or, in bad doctors, of brusqueness and dogmatism. Patients in these circumstances tend to be labelled with a convenient diagnosis which serves to fob them off with a satisfactory explanation, and to touch off a stream of routine therapy.

Hospital practice likewise has its rewarding moments of therapeutic skill, in the accident wards, in the labour wards, in the intensive care units, and sometimes in the skilful handling of modern drugs which can stem the acute infection and the acute depression. But much of the hospital doctor's time is likewise spent in bowing down to the great god Diagnosis. If we are to understand the connection between USE and Disease, we need to be quite clear about what is meant by "a diagnosis", and in the process discover what a disease isn't.

When you walk into a doctor's consulting room, he begins right away to label you. At first his labelling is a rough one—you look "ill". If he is to decide what sort of "ill" you are, he has to examine a few more of your characteristics.

In all of us, at any time, there are a vast number of physiological events taking place. Our kidneys are filtering out fluids, our stomachs and our intestines are setting about any food which comes their way, our hearts are pumping, our senses are sensing, our brains are thinking, and our muscular reflexes are doing their best to handle our postures and our movements, our words and our desires: the body is a complex whole in a state of perpetual process. A vast number of interacting events and interpenetrating processes are going on in all of us all the time—a veritable "flux of events". Your doctor would like to know which of these "events" are not functioning as they should.

The processes which contribute to our organised make-up are interfused and interdependent, but in spite of the interdependence, it is possible to select some of the distinguishable processes and "events" for measurement. By dint of various examinations and laboratory tests it is possible to obtain details of some of the malfunctioning processes and at this point the business of Diagnosis begins. The individual events are collected together into a "bundle of events" and this bundle is the Diagnosis. The doctor is already logically in very deep water indeed.

A doctor learns many things in his medical training. Philosophy is not one of them. He may, it is true, become what is called a philosopher, but this is in the practical sense rather than in the academic sense of the word. Of all stables which currently require a philosophical spring-clean, the medical stable is one of the first on the list. The most appalling logical blunders are being perpetrated daily in the minds and in the consulting rooms of the medical fraternity: and, in our hospital wards and Out-patient departments, all manner of procedures are being carried out in the name of diagnosis. Patients are having needles of various shapes and sizes thrust, with varying degrees of accuracy, into almost every portion of their anatomy.

Why, you may ask, as you lie there writhing, why does this red-faced and perspiring young doctor thrust not once but twice and thrice into my lower back with a thing like a miniature rapier? Why does he say "Sister, Sister, give me a tube", and proceed to withdraw from you fluids which you can ill spare? Why is it important to him? Why is it important to you? And then, as night comes on and you review the situation, a horrible doubt assails you. Does he know why it is important? Does he really understand why he is sticking needles into you?

So next morning you beckon him to your side, and you say, "Doctor, why did you stick that needle in my back yesterday? It wasn't to make me better, was it?"

He will regard you sternly, and if he is talkative, he may say: "We want to diagnose your condition. We want to know what you are a case of. We want to know what disease is causing your pain. The doctor who sent you to hospital thought it might be sciatica, but we want to be sure about the cause. You surely want us to explain your symptoms by finding out what it really is?"

The young doctor spoke frankly and fairly. Diagnosis, he has been taught, is the be-all and end-all of his work. He must affix to each patient a label which will touch off a routine stream of therapy, a stream which is determined by experience of previous patients with similar labels.

You, the patient, have a pain down your thigh: it is a sharp pain, excruciatingly painful, in fact. You want two things. First of all you want the pain to stop and then you want to know what produced it and how to stop it happening again. When you go to your doctor, you are given two things, a treatment and a label. The treatment may be almost anything; the immediate label, in this case, may be "sciatica". You go home with the impression that sciatica has caused your pain. Not only have you "got" sciatica, but sciatica has "got" you.

The power of a word is tremendous. The mysterious quality of illness has made it all too easy for the patient—and often the doctor—to imagine that diseases do actually exist in the way that a table exists.

To Sydenham, in the seventeenth century, each disease was an entity apart from the patient who displayed it. Sydenham viewed symptoms as having behind them some objective entity—something displayed by, but independent of, the patient. This medical fallacy grew strong in the days when medicine was based predominantly on the post-mortem. The post-mortem was, and is, an invaluable method of observing the condition of the dead body and it is quite obvious that certain organs and tissues do show marked changes which must have been present during life. Likewise tissues and tissue fluids which are removed from the living people may show disturbances in their chemical and physical reactions and appearance. In this way, the disease which the patient has "got" comes to be thought of in terms of these changes in organs and tissues and chemical functioning.

The chemical and physical disturbances (events) which a doctor discovers in the body by his various forms of examination and testing are collected together and grouped into a "bundle of events". For example, when you had an acute pain in your leg, your doctor may have discovered that the muscles were wasted and that certain sensations were altered. An X-ray of your back may have shown that certain vertebrae were too close together. These observations of individual "events" were then grouped together into a bundle and given the label "Sciatica" or, more probably, "Prolapsed Intervertebral Disc (P.I.D.)".

It happens to be the case that we cannot in our language refer to events which are going on somewhere without introducing a name, a word, or phrase which appears to stand for something which is producing them. But since most of us still have the primitive belief that to every name an underlying real entity must correspond, it becomes only too easy for the patient to think that an "event" such as muscle-wasting is produced, by some underlying "morbid entity"—by "Sciatica" or the "P.I.D.", even though these are merely *words* used for labelling. The same principle applies to such labels as Multiple Sclerosis, or Schizophrenia or Leukaemia, which the patient is thought to have "got" and which are thought to be responsible for his symptoms. It leads to statements like "We thought he was Schizophrenic, but he is really a case of reactive Depression". There is a parallel to this in the field of Law where the "is it really an X" problem is always coming up: when, to put it in a stylised form, the accepted criteria of applicability of the word "X" are, say, a, b, c, d and e, and then we find a situation in which, a, c, d, and e are present but not b: and we ask misleadingly, "Is it really an X?" In law this question is a request for a decision rather than for a descriptive classification. The question should be "Will it be useful to class it as one?"

This point of view begins to look quite obvious when put in this way and most doctors would reckon to diagnose on the basis of "Will it be useful to label it as one?" But labels have a way of gaining a spurious objectivity, and leading a life of their own.

The bundles of events (diagnostic labels) into which individual events (tissue and chemical changes, etc.) are grouped, do not exist in the same way that the individual events themselves may be said to exist. "Sciatica" does not exist in the way that "muscle-wasting" exists. The process of bundling together events into diagnostic labels is simply a form of classification, adopted from constant usage, embodying the cumulative experience of many doctors. The classifications do not have an independent existence in their own right: they are not there, like the North Pole, waiting to be discovered. They are not "natural" kinds of grouping, in the Platonic sense of "natural" kinds of group, in which things are so-and-so in some absolute sense, so that one point of view, and one only, must be the right one. Leukaemia does not "cause" a raised white cell count: schizophrenia does not "cause" hallucinations. How can a classification "cause" anything except a disturbance in the mind of someone who happens to hear it uttered?

To say this is not to deny the immense usefulness and convenience of a pathological diagnosis. It is simply that diseases are labels—words. There are no diseases—only sick people. What doctors call diseases are states of functioning of many people, no two alike, but similar enough for general concepts to be formed. However, it is only too easy, once doctors have got hold of a recognised disease label, to begin to attribute a false materiality to it—what Whitehead called "the fallacy of misplaced concreteness". By speaking as if disease-labels are morbid entities, doctors are indulging in a sort of medical metaphysics: and all too often they become satisfied merely by pinning on a respectable diagnostic label, a label which satisfies them and their obedient patients.

This obsession with explanatory diagnosis is to a certain extent responsible for the impersonal treatment which patients encounter, whether it be in hospital or from their General Practitioners. The "patient-as-a-person" is easily lost in a too facile labelling process, by which over-importance is given to a "real" pathological lesion which can explain the illness. In the words of T. S. Eliot:

> There is, it seems to me
> At best, only a limited value
> In the knowledge derived from experience.
> The knowledge imposes a pattern, and falsifies.

Bad doctors, who are subsequently shown to have made inaccurate observations on their patients, are doctors who have been determined to make what they find in their patients "match" certain preconceived diagnoses with which they are familiar. They make their perception of the patient fit in with what they think it ought to be. Their knowledge "imposes a pattern, and falsifies". In this, they are often encouraged by patients who will grasp at anything which gives them hope, and which makes them feel that their state is a familiar one which doctors know all about.

Descriptive Falsification

But we are concerned here not with bad, ignorant and hide-bound doctors, who, if pressed, will excuse over-simplification on the grounds of pressure of work. What matters are the good doctors—doctors who with great personal dedication and effort and intelligence are trying to help.

Why is the diagnostic label, as it is used at present, only of limited value and how does it falsify? Consider the case of John G, an insurance broker, aged 42, who was admitted to hospital in the middle of the night with a gastric ulcer which had perforated. He was already known to the hospital and had been treated there four months previously. I only knew him as a social friend and not in a medical capacity, but it was clear that he was creating a tension state in himself which was certain to produce symptoms again of some kind. His work with Lloyds involved sitting cooped-up in a small area in which he was constantly besieged by demands for quotations—demands which needed a very quick response often involving huge sums of money, and requiring the brain of a computer, the tact of a diplomat, and the nerves of a tight-rope walker.

When he had been previously in hospital with a complaint of abdominal pain, he was fully investigated and a barium meal had shown that an ulcer crater was present in his stomach. The diagnosis of "Gastric Ulcer" was made and after a period of treatment with diet and various medicaments, he had lost his pain and returned home: after a short convalescence he returned to his work.

Why was such a diagnosis of "Gastric Ulcer" inadequate, since it was clearly the correct explanation of his pain? Because, when we classify the patient as "a case of gastric ulcer", we are referring not merely to the ulcer crater but to the whole patient, his whole history, his whole personal and social nexus. The patient is not a "gastric ulcer". The diagnosis "gastric ulcer" does not exist in the same sense that the observed ulcer crater might be said to exist. The preoccupation with the label and the ulcer crater dominated John G's doctors to such an extent that they were content simply to have eradicated the ulcer, the pathological explanation of his illness. The fact that John G broke down again, with a perforation 4 months later, shows that the whole concept of Diagnosis had slipped up somewhere. It was not enough to have sent him home with a diet and drugs and a diagnosis which *explained* his ulcer. What was needed was a diagnosis which could *predict* the future course of his illness.

Unfortunately, the word used for a *predictive* diagnostic label is often identical with the word used for an *explanatory* label, and this is usually the name of a specific local lesion or disorder of functioning—"Gastric Ulcer", or "Prolapsed Intervertebral Disc", or "Bronchitis". The muddle arises most insidiously, because the same diagnostic label has to perform the two different roles of explanation and prediction. This muddle is so deep-seated that many doctors will feel in their bones that it is crazy not to admit that "Diabetes" causes a raised blood-sugar, or that "Bronchitis" causes purulent sputum, or that "Depression" causes depression: yet they are not very different from the patient who says, "It's me gastric stomach what's worrying me again, Doctor."

Of course, at first, the patient and his doctor need a satisfactory explanation of the symptom which is troubling him. If a patient starts vomiting, is it due to the lunch-time oysters or is it due to a brain tumour starting to press on centres in the brain? (to take an extreme example). When John G started vomiting blood in the middle of the night, the likely explanation was his old ulcer crater. So far, so good. The ulcer crater is an event, and correct explanation is always in terms of events—oysters, ulcer-craters, intra-cranial pressure. However, once an "event-explanation" has been arrived at, the more difficult matter then arises of constructing the predictive diagnosis which will suggest ways of handling the patient in the future. Explanation and prediction are two quite separate matters.

This means that a predictive label must include more events

than those needed for the immediate explanation. Lack of amines in the brain may, of course, sometimes produce depression, but the predictive-label "Depression" must include many more alterable events than amine-lack. Lack of insulin may, of course, produce a high blood-sugar, but the predictive-label "Diabetes" must include more alterable events than insulin-lack. Bronchial inflammation (bronchitis) may of course produce purulent sputum, but the predictive-label "Bronchitis" must include many more alterable events than bronchial inflammation. The label "Gastric Ulcer" must include many more alterable events than the actual gastric ulceration; and so on.

Medicine is just about the only science left which still clings to hopes of finding a "true explanation". We should label to predict, and not merely to give a "true explanation" of what is causing the trouble.

If I see an animal in a field, I may label it "dangerous bull" and modify my action accordingly, whilst the farmer labels it as "my best Shorthorn", and his dealings with it differ from mine: his is not necessarily a truer view of the animal. Likewise, John G in the ward may be to the house physician "a case of gastric ulcer", to the night nurse "two Soneryls at 10", to the hospital secretary "bed 22 occupied", to the hospital newsvendor "a *Daily Mail*", whilst the poet in the next bed may be thinking "Pale death knocks impartially, Pallida mors impar", and making a predictive diagnosis, almost certainly wrong, that pale death is about to knock. All of these people adopt labels which may help their view of John G during his hospital stay, and which predict what may happen in their relationship to him over that relatively short period.

We can improve the predictive value of our diagnostic label by learning to include more and more events in the "bundle of events" which makes it up. The events can be collected at all levels from the patient—physical, chemical, bacteriological, cells, tissues, organs, and so on up to the higher mental functions and to his personal and social behaviour. It is not that events at the physical and chemical levels are to be seen as an *explanation* of events at another "higher" level, but simply that, when we come to apply a diagnostic label, we should make up bundles of events from whatever levels may help us to predict the patient's future and to handle him accordingly.

Description and Prescription

One event which had not been observed in John G's case was his habitually tense USE. This event should always be considered when making a selection of events to be included in the "bundle of events" which makes up a diagnosis. Any diagnosis which has not considered it is liable to fall down—sometimes to a lesser, usually to a greater degree. John G's ulcer might well not have perforated if, during his first hospital stay, his USE had been noticed and tackled.

If, then, USE is to be included in the making of a diagnosis, we must realise that a diagnosis should encompass two things. It should contain a *description* and it should contain a *prescription* —a description of events which are malfunctioning and a prescription of principles for future amelioration and prevention.

Diagnosis has for too long only looked for descriptive explanations (whether they be "dispositional" explanations, "efficient-cause" explanations, or "final-cause" explanations).* Such descriptive-diagnosis may satisfy the desire for pigeonholing, but it does not necessarily carry implications for long-term prevention. A descriptive-diagnosis tends to see the patient in a deterministic manner, at the mercy of his genes, his chemistry, his reflexes and his social commitments. Patients tend to be seen as objects lacking free-will, and not as persons. In the preoccupation with pathological minutiae, the patient-as-a-person is lost.

This "descriptivism" (as it is termed by the philosophers)[19] can, of course, be counter-balanced, and in many doctors it is counter-balanced, by art and understanding: but it often happens that doctors have less intuitive understanding than ordinary unschooled people. From his earliest days in the dissecting room, the medical student has to stifle troublesome feelings which may be stirred up, say, by handling a corpse. He soon learns, in this and many other situations, to filter out his ordinary feelings and to lock himself in a scientific enclosure in which he denies himself the ordinary reactions of the outside world. By so doing, he is able to see his patients as objects for scientific scrutiny, and not personally in a way which might require over-involvement. Such an approach may give him

* See Chapter 7, p. 111.

confidence and a tranquil mind, but it is a confidence purchased at the price of great psychological blindness.

Such *descriptivism* is naïvely deterministic. It does not take sufficient account of the patient's freewill. *Prescriptivism*, on the contrary, says that, as long as people are willing to make up their minds about what to do next, they can be free. But in making up their minds about future action, they have to make a prediction about the future—an opinion, a hypothesis, a diagnosis of what seems likely to happen or desirable to make happen. Such a "prescriptive-diagnosis" does not imply that all future events are or could be predictable. In the realm of health, it simply seeks to include in its diagnostic "bundle" such events as seem likely to affect the patient's health in the future. It seeks not simply to describe the present, but to prescribe action for the future. And it prescribes such action on the basis of principles, of priorities.

The Alexander Principle says that USE will always affect FUNCTIONING. It says that USE should always be included in a prescriptive diagnosis. It says that it will always be important for anyone anywhere to know how to sustain a good manner of USE, in sickness or in health. It seeks to replace the concept of CURE by the concept of PREVENTION.

Preventive Medicine

From the patient's point of view, all medicine is preventive medicine. Preventing what it is feared might happen as well as what is actually happening. Preventing pain, preventing a lump from growing and spreading: preventing headache, insomnia, agitation, depression: preventing cough, fatigue, giddiness: preventing an arm or leg from behaving abnormally: stopping overweight, stopping underweight: preventing the hazards of childbirth, preventing even the child being born: preventing the aching back, the tingling fingers, and so on.

Health, for most of us, implies that we are able to do, without strain, the things we expect to be able to do: and that we are able to *be* without strain the way we like to *be*. Restoration to contented functioning is what most people mean by being healthy. But when we come to worry about the presence of possible *disease*, we are concerned in the main about *what may happen next*. We are concerned to prevent future trouble either in the immediate or distant future.

The general scientific and technological explosion which got under way in the 1920's has dominated medical thought and training for the past half century. But the explosion is now becoming a spent force. The gains are immense: the new medical skills and medical tools have undeniably altered our whole concept of how to deal with severe illness. Many conditions whose progress could not be adequately prevented before can now be checked. But in the words of Professor Sir Macfarlane Burnet OM, FRS:

"The contribution of laboratory science to Medicine has virtually come to an end: almost none of modern basic research in the medical sciences has any direct or indirect bearing on the *prevention* of disease or the improvement of medical care."

Most readers of this book will know that this is true. They know that they have good reason to thank Medicine for taking the terror out of acute illness: but they also know that it is rare nowadays to find a really healthy happy human being—and one who sustains such health over most of a lifetime. They accept that many symptoms can be managed by the use of powerful psychotropic drugs or by powerful antibiotics, but they are also aware that medicine has little to offer them in certain other conditions in which their functioning is not as good as they might like it to be.

The Duke of Edinburgh, speaking to the British Medical Association, had something to say about this:

> Much of the progress in medicine has inevitably been made at the price of deeper and narrower specialization, but the individual is still one unit, and so far as his personal health is concerned a unique unit at that. I am all for studying bits and pieces, but I hope that treatment will remain directed to the whole. The . . . cure of disease is a laudable object but it can be followed blindly. Medical science has to face the fact that remedies for one problem may give rise to others. Of one thing I am quite sure, and that is that the Common Health is more than figures showing improved birth rates, death rates, and the incidence of disease.

The British Journal of Hospital Medicine makes the same point (November 1971):

> Advances have brought most acute infections under control but the frequency of chronic disease has increased in both relative and absolute terms . . . The doctor's new role is often an uncomfortable one, since he must continue to care for patients whose disease he cannot cure . . . The sense of responsibility has to be extended from care of the sick to care also for the *potentially* sick.

The Use—Diagnosis

If I am right—and I think I am—that USE is the single most important factor which remains to be dealt with by medical science, one reason for the lack of help which the average person gets from the average doctor in certain cases is apparent. *Doctors—through no fault of their own—have not been trained to observe in detail the variegated patterns of mis-use which are going on in all of their patients.*

To say all of this is not to claim that the faulty manner of use is the main cause of most unexplained malfunctioning (although in many cases there is a clear causal connection): it is simply to say that although most other types of diseased functioning come under the scrutiny of the medical eye, USE is usually ignored: it is not one of the "events" which is considered when putting together the diagnostic "bundle-of-events". And since those whose USE is reasonably good seem to avoid or at least postpone many of the ills to which human beings are prone, any diagnosis which does not consider this factor is incomplete. Fundamental to the preventive care of a patient s the teaching of improved habits of life, and, at core, this means teaching an improved USE.

Such a conception goes right beyond our present regimes of medical attention. It in no way precludes the need for an accurate pathological descriptive-diagnosis: but it means that no prescription for treatment is complete until a person's USE has been taken into account.

Chapter 6

USE AND DISEASE

IF USE IS indeed the single most important factor which remains to be dealt with by medical science, then it would be reasonable to expect that its relevance would be most obvious in those medical conditions which, excluding minor coughs and colds, are far and away the greatest cause of general ill-health in our present society.

Two types of disorder stand at the top of all the illness league tables—mental disorders and rheumatic disorders. If we consider only the working population (excluding adolescents, housewives, the elderly and chronic disabled who contribute to the bulk of mental disorder), we find that 36 million working days are lost each year from mental disorders. The rheumatic disorders are responsible for almost as many—35 million days lost by the working population each year. The relationship between USE and Mental Health will be taken up in the next chapter, but in this chapter the relevance of USE to certain physical disorders—in particular the rheumatic disorders—will be considered.

Rheumatism

The term "rheumatism" is a rag-bag which includes a vastly differing range of conditions. It was introduced in medieval times by Galen from the Greek word "rheo" (flow), at a time when medicine believed in the four "humours" of the body whose "flow" was liable to be deranged: in rheumatism, through "acrimony of the humours", abnormal flow was thought to take place into various body cavities: in gout, there were thought to be abnormal drops of humour (guttae) in the joints. It became the practice to refer to any sort of shifting pain as "rheumatism", although in the eighteenth century the term was applied mainly to the muscles. It was recognised that the joints could be involved, but secondarily to a muscular disorder.

In 1827 Scudamore, in a comprehensive treatise on rheumatism, thought that the fleshy parts of muscle were not affected, but that the pain came from tendons and their fibrous insertions into bone. In 1904 Gowers[20] thought that muscular pain came from what he called "fibrositis"—a local inflammation of muscle, although other workers blamed either rigidity or weakness of muscle. At this time Alexander began to expound his concept of mis-use, in which he proposed that attention should be paid to the *general* muscular co-ordination, and not simply to the local site of the pain in the muscle or joint. He also suggested—in majestic language—that the basic fault was psychophysical, and that it lay in "faulty preconceived ideas", "debauched kineasthesia", "inaccurate sensory appreciation".

His ideas received support when, in 1937, Halliday[21] found that of 145 cases of rheumatism, 33% were psycho-neurotic as well: of the cases which went on for more than 2 months, 60% were considered psycho-neurotic. This was confirmed by several other leading rheumatologists—Ellman[22] in 1942 suggested that "the muscles serve as a means of defence and attack in the struggle for existence: if the external expression of aggressiveness is inhibited, muscular tension may result which is felt as pain": and he found that in 50 cases of muscular rheumatism, 70% suffered from psychological disorders.

As electromyography came to be used, many people showed (Hench[23] 1946) that "psychogenic rheumatism is one of the commonest causes of generalised or localised aches and pains". I stressed at this time that we should not use the term "psychogenic": that it was not a question of the "psyche" generating disorder in the body, or the body generating disorder in the "psyche", but rather that the rheumatism and the psychoneurosis are *both* manifestations of an underlying failure to achieve a balanced "resting state" of USE after stress.

Subsequently I published a study of students[24] in which the 10% who complained of persistent muscular pain all manifested a severely disturbed postural balance: and Eysenck,[25] in his book *The Dimensions of Personality*, found a definite and fundamental correlation between neuroticism and an unbalanced postural balance. The "neurotic" will invariably be found to have poorly balanced USE, and the worse the imbalance the greater the likelihood of pain.

I suggested at that time that we should not talk about Misuse as being a psychosomatic disorder but rather that it would be considered as a "stress disorder": a stress disorder being one which habitually involves bodily systems beyond the relevant ones, and in which the organism does not return to a balanced resting state after activity. I suggested that bodily systems could be involved at four levels:

1. Physiological changes
2. Emotional changes
3. Behavioural changes
4. Structural changes

and that all levels might be affected at the same time in the composite disorder of "Mis-use". "Rheumatism" in fact, as I saw it (and still see it), should never be considered simply in terms of the local part of the body which is malfunctioning, but should be taken in the context of the mis-use of the whole body, involving as it does, physiological changes, emotional changes, behavioural changes and structural changes.

Part and parcel of this "USE" concept of rheumatism (and indeed of many other bodily disorders) is my disbelief in the "wisdom of the body". The body is not wise: it is frequently stupid. Perhaps I might digress for a moment to stress what I mean by its stupidity, and how it is that the body may be "wise" about *ends* but not about *means*.

The Stupidity of the Body

For a human being to remain alive, certain "variables" must remain within definite limits. Many variables in our bodies—the length of our hair or the length of our nails—are not essential to life (except in a social sense): but the temperature and acidity of our blood, the amount of oxygen, sugar, salt, protein, fat and calcium are of vital importance and a matter of life and death. They are kept constant by a ceaseless interplay of adjustments. Even if we don't drink for three days, the amount of water in the blood will change very little. If we then drink six quarts of water in six hours the blood volume still won't change much, although the kidneys have to work overtime to pour it out into the bladder. These mechanisms are on a stimulus/response basis. If there is too much swing away from the desired norm, the body (in its stimulus/response "wisdom") is stimulated to use one of its many systems to restore the balance. If its systems cannot come up trumps, it searches its environment for the necessary constituents—salt for the sweating miner, heroin for the junkie's transient "steady" state, alcohol for the liver which can no longer convert other food properly, nicotine for the quick energy of sugar release, sexual discharge for the irritable restless gonads, colour and music to restore momentary peace to the restless brain.

This might seem like the wisdom of the body, but it is wisdom at a low level. It is the wisdom of end-gaining.

Consider another "essential variable"—your blood pressure. If by some mischance the part of your brain which usually regulates it is knocked out, another part of the brain will take over the job. If this part is knocked out, various ganglia outside the brain take over. If these ganglia are knocked out, the blood vessels themselves attempt, by contracting or enlarging, to regulate the pressure of the blood they contain.

The body, in fact, usually has several alternative ways of doing things. In carpentry terms, it is a bodger. An artful bodger and an end-gaining bodger. There are certain things it wants—so much sugar in the blood, so much salt, so much oxygen, so much protein, such and such a familiar sense of muscular equilibrium, such and such a state of mental calm—and if its easiest ways of getting them don't work, it bodges around with its other systems or muscles to knock something together which will do, until finally, it runs out of alternatives and begins to seize up. This is "end-gaining"—the determination to get short-term ends on a reflex stimulus/response basis, without ensuring that there are no harmful by-products.

Reflex end-gaining does not pause to see whether the alternatives will be constructive or destructive in the long run. Such destructive alternativism is not "wisdom". The badly burned body will recklessly pour out its body fluids through the burnt surface until it dies from fluid loss. The asthmatic in his end-gaining anxiety to take in more and more air will use muscles in the upper chest in such a way that he cannot release them properly to let the used air out.

Bodging, which uses potentially destructive procedures, is always a risky business, although much of our fortuitous evolution to date has been a case of "artful bodging". We ourselves have it in us to decide our future in the spheres which we consider important. With all our short-comings and conflicting systems and desires, we can still act by deliberate intention, exercising our mind. Instead of being at the mercy of a bodgedup system of physico-chemical reactions or muscular reflexes, we have it in us to become self-adjusting. We have it in us to live by principles which we ourselves have personally selected—a life of "constructive alternativism", using a new body-construct, rather than a life of destructive alternativism.

Use and Rheumatism

Nowhere is the "stupidity" of the body more apparent than in the case of the rheumatic disorders. Let us take as an example the condition of osteo-arthritis of the hip-joint. I am not here concerned with the factors which may have led to its development—although I suggest that mis-use is one of the most important—but with what happens when the condition is beginning to show itself. The earliest sign is a narrowing of the width of the joint, a narrowing which leads to a shortening of the distance between the hip and the ground. This shortening usually leads to a further faulty distribution of body weight so that it is mainly carried through the affected hip. The result of this wrong distribution of weight is that the arthritic condition progresses and further shortening takes place on the affected side: this leads to putting further weight on the affected hip-joint, and so on, until the familiar picture of distortion and leg-shortening presents itself.

Such arthritis of the hip cannot be considered purely as a local event, although, of course, the point may come when local surgery is needed in dealing with the local condition. From a prevention or rehabilitation point of view, the use of the whole body has to be considered. When the early condition is starting to show itself, attention must be given to the *general* use, to see that the balance is not thrown more and more on to the affected side.

This principle applies equally to such relatively slight but infuriatingly persistent conditions as "tennis elbow", painful stiff shoulders ("frozen shoulder"), and minor aches and pains in neck, chest, back and legs. Such conditions may certainly disappear after rest or rubbing or injection or physiotherapy. But often they persist and become a source of worry and dissatisfaction with medical care. More often than not, such conditions will be associated with a small but un-diagnosed disorder of the general USE—plate 17 shows a young girl with just such a persistent leg pain whose tendency to stand with her weight wrongly distributed was only noticed when she was accurately photographed against a grid.

But these are usually small pains and pin-pricks. Of far greater importance are the major USE-disorders of the spinal column. The need for good USE-training in the various postural deformities of the spine—kyphosis, lordosis and scoliosis—is apparent; and even in spines which are severely affected, as in polio or idiopathic scoliosis, a great deal of help can usually be given to correct some of the deformity and to prevent its increase. But by far the greatest number of patients seen in a Rheumatology Clinic (excluding the trivia, and arthritis affecting the limbs) will be one of two conditions. The first is what is known as cervical spondylosis; the second is the almost pandemic pain-in-the-back. These conditions are so prevalent and often so unresponsive to medical treatment that they each need special consideration here, since it is in these two conditions that the Alexander Principle has much to offer.

Cervical Spondylosis

Since the head-neck balance was thought by Alexander to be the primary seat of mis-use, he had much to say about symptoms which arise from such mis-use. The varieties of mis-use in the head-neck region are infinitely complex. If we look at only a tiny sample of mis-used necks (plates 18a, b, c, d, e) we see how much they vary: and yet an X-ray report on them never mentions even the *presence* of mis-use, much less the *type* of mis-use. Added to which, a "still" picture of a neck gives only slight indication of the nuances of muscular usage which are going on all the time, whether they be produced by speech, or swallowing, or gesturing, or emoting or searching the surroundings with the eyes or the ears.

The commonest of these mis-uses involves a drop forward of the middle of the neck, and this is accompanied by a pulling of the head back on the top of the neck. Plates 18a, 18b and 18c give examples of this: plate 18d shows the same thing, although in time the neck has become so collapsed into the hump that it may not at once be obvious just how much collapse has taken place.

This collapse will involve a tension of the muscles which go up into the back of the skull—not only the big external muscles (fig. 17a, b, c) but also the smaller internal sub-occipital

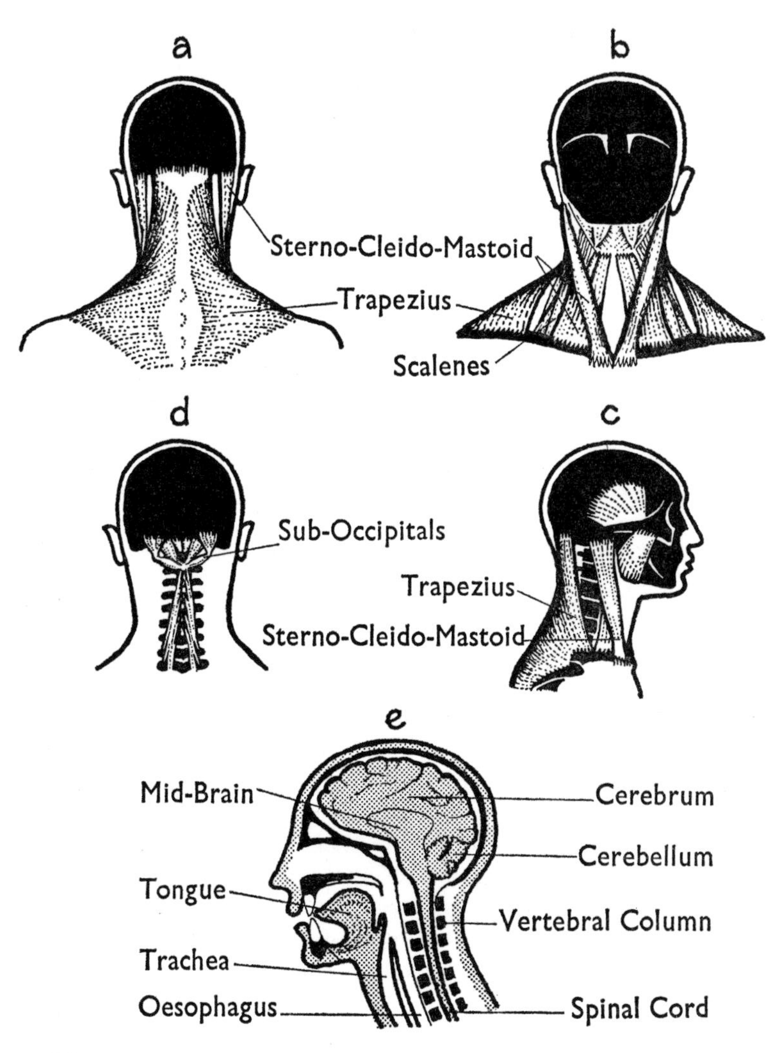

Figure 17

Only the external and internal muscles of the neck are depicted—certain intermediary muscles are omitted. The sub-occipital muscles play a part in fine skull movements. The other muscles which relate the skull to the neck are mainly inserted into the back of the skull. There is great disparity in strength between the anterior (flexor) muscles and the posterior (extensor) muscles. In the upright posture, strong extensors encourage pulling-back of the head. To carry out relatively automatic functions of mastication, swallowing, breathing and phonation whilst the body is held erect is a complex task. (See Raymond Dart: *The Attainment of Poise*. S. Afr. Med. J., 1947, 21, 74.)

muscles (fig. 17d) which are placed around the junction of the neck and head. Much was—ill-advisedly—made of these small muscles by some of Alexander's earlier medical supporters who encouraged him in a belief in an almost magical potentiality to be obtained from releasing such muscles so as to give full range to his "Primary Control". One doctor wrote "The primary relation upon which all more ultimate relations depend is that relation established by the small group of muscles which comprise the atlas-occipital, axis-occipital, atlas-axis system. The stupendous importance of this relation in the functioning of muscles cannot be realised by a mere description of its existence."

Here is another example of the "stupendous" atmosphere which dogged Alexander in his life-time and which masked the actual factual analysis of the immensely varied forms of misuse which affect so many necks: it is not just *one* type of muscle-balance which is at fault, but a complicated network of muscle-pulls throughout the neck.

Sometimes, instead of a collapsing of the neck forward, there will be an over-straightening of the neck with slight movement of the body of one of the neck vertebrae backwards on the one below it, as in plate 18e. These cases, in my experience, are usually associated with really intractable neck, head or facial pain.

If we look at the neck from the back, we will often see a slight curvature sideways at the base of the neck, in the hump (see plate 22). This is usually ignored in X-ray reports, but, again, it is often associated with intractable symptoms. If you study the back of the neck, as in plate 12c, you will notice two tense contracted muscles just where they insert into the skull. When there is a sideways twist at the base of the neck, one of these two muscles will be more prominent than the other and the resultant tension imbalance is even harder to release than the more symmetrical type of faulty tension.

These and many more types of mis-use will always be found in the condition of cervical spondylosis (head and neck arthritis) or cervical disc pressure. These conditions affect 85% of us in our 50's, and most doctors will agree that they are difficult to treat and are long lasting.

The usual symptom, at first, is numbness or tingling in the fingers of one hand, due to pressure on nerve-roots from the compressed and fore-shortened neck vertebrae. Sometimes the first symptom is severe pain in the neck, shoulder or arm, which typically is worse in the early hours of the morning, when the warmth of bedclothes has perhaps produced a congestion in the already narrowed passages which carry nerves outward from the spinal cord. Sufferers find that they have to get up, or place their arm outside the bedclothes, or in some new position to relieve the pain.

Re-education of the faulty USE which accompanies cervical spondylosis will usually produce a relief of symptoms.

Pain in the Back

Over half the adult population experience severe lower back and sciatic pain. Of the £190 million estimated annual cost of the rheumatic disorders, a high and increasing proportion are due to back pain. Middle-aged workers with back trouble stay away from work for an average of five months, and since a large number of them will be off work for less time than this, it means, on the other hand, that a sizeable proportion will be away for more than a year.

In most cases there may be some obvious immediate cause for the acute pain—a prolapsed disc or ankylosing spondylitis: but most of the people with painful backs will find themselves given a diagnostic label such as lumbo-sacral strain, sacro iliac strain, postural backache, ligamentous strain, fibrositis; plus, from osteopaths, sophisticated labels like "facet-block" or the unsophisticated label of an "osteopathic lesion".

Over 1,000 possible reasons for back pain are cited in one American text book. USE is not to be found amongst them. The failure to cope medically with back pain is exactly what might be expected where there is widespread ignorance of what constitutes a properly used back. My own experience, *quite categorically*, is that most forms of back pain, even after there has been unsuccessful surgery, are best treated by making USE re-education a prime necessity. *Of course*, there may be alleviation by manipulation, or physiotherapy, injection, traction, corsets and so on: but the basic problem remains, that these procedures do not alter the general manner of USE, and this general manner of USE constitutes a Damoclean sword which is liable to fall when external or internal stress builds up.

Only too frequently such disabling back pain attacks men

and women of very high calibre who are making a big contribution in their chosen sphere. Often they will be under thirty years of age, and they face a prospect of crippling pain which will impose a restrictive pattern on their lives, whether it be in a rearing of their children or in the development of their working life. Eventually, after casting around for any conceivable therapy that might help, they resign themselves to a greatly restricted existence, propped up by corsets and aspirins and loving relations and friends: and very often they become profoundly depressed, since the mis-use pattern which led to their back trouble will, more often than not, contain a depressive-slump element in its components.

It cannot be emphasised too strongly, so I will write it in capitals. IT IS WRONG TO TREAT A PAINFUL BACK AS A LOCAL CONDITION. BACK PAIN IS ALWAYS ACCOMPANIED AND PRECEDED BY GENERAL MIS-USE.

This general mis-use will have many variations. But almost always three things can be seen to be wrong. The thorax (chest cage) will be thrown over to one side (and may be slightly rotated on the lumbar spine) (plate 19). The shoulder-blades will be raised, so that much of the muscular supporting work of the back is being done by the shoulders instead of the mid-back (plate 19). And the breathing pattern will be seen to involve a slight arching forward of the lumbar spine and the predominant habit of breathing-in with the front of the chest and the abdomen. There will be many other associated tensions, depending on the type of pain. An acute disc lesion will probably show flattening and rigidity, a chronic back may be hyper-lordotic. But, strange as it may at first-sight appear, a painful lower back should always first be dealt with by sorting out the mis-use patterns in the Neck and Hump, since only when an improved co-ordination can be maintained in the upper part of the back, can there be an improved USE in the lower back and legs.

Arthritis

Before leaving this discussion of rheumatic disorders, a mention must be made of the condition of rheumatoid arthritis. It differs from osteo-arthritis in that it is a general joint disease. It affects at least 1% of all males and 3% of all females: in mild cases there is little disability but severe cases show advanced crippling and deterioration of many joints and tendons. It is another example of the "stupidity" of the body in that the "adaptive" response of the body is a major cause of the disease.

A study of 400 cases in 1952 by the social workers of Peto Place (now the Arthur Stanley Institute) left no doubt that in at least 60% of cases of rheumatoid arthritis, there had been a preceding stress. Over 95% of a series of cases which I reported at the Royal Society of Medicine in 1964 occurred after what I considered to be preceding stress.

A typical history, as written by one of my patients, was:

> My mother was taken to hospital with her second stroke aged 73: I went backwards and forwards for 16 months to Mile End, sometimes at 10 o'clock at night, sometimes 8 o'clock in the morning, sitting with her for hours, then rushing home and on to work. After she died I woke one morning, my feet felt funny and couldn't walk. I thought it was chilblains. It got worse till I had to borrow shoes too big for me, then I came on to you.

Or another one:

> We were living under very crowded conditions for five to six years, in premises that had been condemned by the Council due to overcrowding.

Or again, from an intelligent librarian:

> It became necessary to assume the responsibility of providing and sharing a house with a parent, thereby changing my way of life. This venture had to be undertaken on a strictly limited income, and making ends meet gave rise to anxiety. The curtailment of my personal freedom was realised with regret. The situation was accepted, there being no question of escape, rather of anxiety to fulfil duties.

I doubt if anyone who has worked in this field would wish to deny this stress factor, but one is still faced with the question of why one person reacts with rheumatoid arthritis and another person doesn't: indeed, the same question applies to most illnesses—most of us can cite quite appalling psychological traumata or physical strains in our youth and maturity, which we weather in one way or another. Stress, in the event, seems to depend on what seems like a stress to any one of us: in other words, on our body construct, the way we "construe" our situation, and the way we "structure" our response to it through our manner of USE.

"Rheumatoid" patients exhibit in their USE a degree of muscular agitation which is not present in other rheumatic patients. I first noticed this in relation to excessive head movements made during speaking, but closer analysis showed a wriggling movement not only in the muscles of the spine but also in the limbs. Rheumatoid patients in the early stages of their condition, generally tend to be "wrigglers", in a state of muscular agitation: and if they are not, it is either because the pathology has led to fixation, or because pain and discomfort have led them to keep more still, or because they seek to control their agitation by excessive tension. It is particularly in the communication-situation that this muscular agitation becomes apparent and it may serve as an early diagnostic pointer.

Many of them are shy, sensitive people whose feelings are easily hurt—people with difficulty in communicating. This difficulty might account for the meticulous, orderly conscientious life which they like to lead, since they like an environment in which learned stereotyped responses will suffice. It has struck me when dealing with them that through this feeling of inadequacy, in the communication situation, they make exaggerated muscular agitation. Many of them, by a sort of puppy-dog wriggling, put up a false front of niceness which belies their preferred disposition. It does not seem to me outside the bounds of possibility that this constant muscular restlessness may lead to joint dysfunction: and it cannot but be helpful to teach them a muscular resting-state in which their exaggerated homeostatic swings are damped down.

Physical Medicine and Physiotherapy

Most hospital Rheumatology departments in the past combined their function with what is known as "Physical Medicine"—that is to say, the use of *physical* methods as opposed to *chemical* methods of treatment. This sharing of functions has grown up from the fact that Rheumatologists share with Physical Medicine specialists the need for the use of physiotherapists, trained in such physical methods. Nowadays most Physical Medicine specialists prefer to think of themselves as predominantly Rheumatologists, mainly because of the link which this gives them with general medicine, and also because the physiotherapist, who carries out the actual day-to-day treatment on the patient, comes to know a great deal more about the vagaries of muscle than they do.

Much of a physiotherapist's time—when it is not used in various methods involving electricity—is spent on the educating and re-educating of muscle. A major part of their time will be spent on the various rheumatic disorders but they will also be involved in exercises of various kinds—pre-natal and post-natal exercises, breathing exercises, mobilising and strengthening exercises for the back, mobilising fixed shoulders and knees and hips, correcting such postural defects as flat feet, knock-knees, winged shoulders, lordotic backs. Much time is spent in the after-care of people who have broken their bones, suffered amputation or nerve injury, or who have undergone serious orthopaedic surgery to backs or hips or other regions: both before and after an operation patients in the wards are helped to become active sooner than they might if left to their own desires. Much time is spent on the rehabilitation of patients who have had a stroke or suffer from long-lasting nervous diseases, like multiple sclerosis or cerebral palsy. And, of course, a major part of the time is spent on "placebo" therapy in which the friendliness of the physiotherapist and the comfort of massage and heat and so on, is a major part of the treatment. The varieties of procedures are immense but unfortunately they rarely include a knowledge of Alexander's work, which throws a completely new light on exercise-procedures.

Many physiotherapy departments also have personnel trained in manipulative procedures which used to be the preserve of the osteopath and the chiropractor. And many of them use relatively gross manipulative methods in the form of traction to the neck or the spine or the limbs. Such manipulative methods are often a feature of "health farms" which have taken over much of the work that used to be done in the past by spas. In the old days the spas provided an amiable outlet for the wealthy to spend their cash in pleasant surroundings. Now that the spas have been nationalised in Britain, the "health farms" likewise provide a pleasant health-giving holiday for the wealthy or hypochrondriacal, and enable various forms of physiotherapy, etc., to be carried out in pleasant surroundings.

Very few physiotherapists in the NHS or in private practice, have at present an understanding of Alexander's conception of USE. Alexander himself attacked them ferociously, since much of what they did was, in his opinion, based on the "end-gaining" principle and they did not seem to understand the need to treat *local* lesions on a *general* basis.

Physiotherapists have been slower to learn Alexander's ideas than the modern physical educationists. There is no doubt that before long the pressure of demand from patients will make it necessary for physiotherapy training schools to wake up to these ideas.

In all the types of situation mentioned above in which physiotherapy is given—the rheumatic disorders, the spinal disorders, postural disorders, neurological disorders, breathing disorders, ante-natal and post-natal care, general ward care and rehabilitation, the Alexander Principle has much to offer. The only limiting factor is availability of time and the capacity of the sick or elderly patient to co-operate. Restoration to the type of adequate functioning which is envisaged in this book is not at present achieved by our present hospital care, either in the wards or under out-patient care. To say this is not to decry the patient and devoted work which the physiotherapists and occupational therapists are carrying out: it is simply that, as things are at present, there is a gap in their training and in their working procedures.

Breathing Disorders

Nowhere is this more apparent than in the treatment of breathing disorders. To take one obvious example, asthma deaths are increasing, in spite of modern drugs which can counteract the acute attack. It is no good simply blaming increased environmental stress, mites in the house dust, or the increased use of steroid drugs and inhalers which give temporary relief. Something is still missing from the story: and, as usual, this something is the "use"-explanation, which tends to be overlooked. *The asthmatic needs to be taught how to stop his wrong way of breathing.* Breathing exercises have, of course, frequently been given by physiotherapists for this and for other breathing conditions but the fact is that breathing exercises do not help the asthmatic greatly—in fact, recent studies show that after a course of "breathing exercises", the majority of people breathe less efficiently than they did before they started them.

There is no shortage of information about the *physiology* of breathing—most of us know that too little oxygen or too much carbon dioxide will make us want more air and we know that various reflex mechanisms in the brain, blood vessels and lung will work automatically to keep the breathing process going. This is what starts happening when we are born. This is what stops happening when we die. But this physiological account of reflex breathing does not tell us much about *how to breathe.* It is not only in the sphere of medicine that there is this lack of knowledge. Actors, singers, speech teachers, and speech therapists have a special need to know about breathing, as, of course, have all teachers of physical education. Yet in all of these fields —medicine, communications and physical education—there is a paucity of information about wrong breathing habits. The asthmatic does not need breathing exercises—he needs breathing *education.* He needs a minute analysis of his faulty breathing habits and clear instruction on how to replace them by an improved use of his chest. Such chest-use cannot be separated out from a consideration of the general manner of use. *All* patients with chronic bronchitis and asthma have a significantly high score in the Eysenck personality inventory, and the Cattell self-analysis test. These personality disorders increase as the chest trouble increases: and to repeat it again, there is a very high correlation between personality disorder and mis-use.

I have for some years given some very simple instructions to patients about their breathing, and these are reprinted in Chapter 11. As with all such written instructions, however, five minutes practical instruction is worth any amount of reading about it.

Stress Diseases

I have so far mentioned in this chapter many of the types of medical condition in which the Alexander Principle is most obviously needed. To this list should be added the diseases of civilisation—the so-called "stress" diseases. Whilst one would hope that knowledge of the Principle will lessen the incidence of these conditions, for the most part its role will be in the lives of people who have already acquired a "stress" condition and who are needing to change their whole *modus vivendi*.

High on the list is Hypertension—raising of the blood pressure to the point at which there is a risk of cardiac damage or stroke. It is well known that emotional factors play a large part in raising the blood pressure: frequently a patient's pressure will fall considerably when he is resting in hospital or simply when he is less afraid of the medical situation in which he finds himself. I have found blood pressure to drop by as much as 30 points after a half-hour re-educational session in which tense muscles were relaxed: and it seems reasonable to suggest that since most blood vessels traverse or are surrounded by muscles, any over-contraction of the muscle is bound to squeeze the lumen of the blood vessel and thereby make it more difficult for the blood to be pumped through them by the heart. The less the obstruction to the blood flow, the less the pressure.

I see a good number of people who have had a coronary thrombosis. I have never yet seen a case in which the upper chest was not markedly raised and over-contracted. The "powerful" tycoon impression is often accompanied by the blown up, over-filled chest. I regard it as essential that such patients should be taught to release their chest tension and to do so in a way that is accompanied by an improvement in their general USE.

Gastro-intestinal conditions figure high in the list of stress conditions, whether it be gastric and duodenal disorders, "spastic" colon, ulcerative colitis, rectal spasm, or anorexia nervosa in the young. Along with these there are less easily defined symptoms of abdominal discomfort—compulsive air-swallowing, abdominal bloating and belching, and just simple constipation. And there is, of course, the undiagnosed pain in the stomach. The tenth most common cause of admission to hospital in males (and the sixth most common in females) is abdominal pain which remains unexplained after investigation. In children and adolescents it tends to be low down on the right side: in middle-aged people it tends to be higher up in the centre. Children do not grow out of these pains—they have been shown still to be present twenty years later. Many of these patients have their appendix taken out but this does not diminish the re-admission rate to hospital, which is high. It stands to reason that the fullest investigation must be carried out to see if there is pathological change: but where full investigation has been completed and treatment has not resolved the condition, it is often useful to tackle the mis-use which invariably accompanies these conditions. These patients with unexplained abdominal pain also have a high Eysenck neuroticism score—and this, as we know, correlates closely with postural imbalance. More often than not these patients will be found to have a slight sideways displacement of the thorax on the lower back and often a rotary twist of the dorso-lumbar spine with associated muscle spasm. This should always be looked for in cases of unexplained abdominal discomfort.

I am often asked to see migraine sufferers and it is rare to find one who cannot be helped by learning to release faulty tension around the head, neck and face. Whilst there may well be a constitutional predisposition to migraine, in the form of arterial spasm which is hormonally precipitated, there is nevertheless always an added factor of excessive neck tension, which perpetuates a condition, often for days, which should resolve in a few hours. This has been realised by Migraine Clinics, but owing to their lack of an understanding of the complexity of the muscular tensions in the neck and "hump", they have not been very successful in their relaxation therapy.

I see a number of epileptics and petit-mal sufferers, and, provided they are willing to undergo re-education—many of them have a temperament which is impatient unless they are filled up with drugs—it is often possible to reduce markedly the incidence of attacks, to the point of complete disappearance in a number of cases.

A certain number of gynaecological conditions can be helped by the USE approach: dysmenorrhoea, retroversion of the uterus and vaginismus can be helped. The application of the Principle to sexual functioning will be discussed in detail in Chapter 8. A knowledge of how to maintain a stable integrated pattern of USE is also invaluable at the menopause—a time when the awareness of the postural model (the body construct) may often become disturbed, with resultant feelings of unreality and depersonalisation.

Perhaps the most obvious field for application is in the therapy of the various muscular tics and cramps, which can vary from the "occupational palsies" such as writers' and telephonists' cramp, to severe conditions like spasmodic torticollis and persistent spasm of the shoulders and trunk. I should mention here the condition of "atypical facial pain"—acute pain, usually across the nose, cheek and eyes, differing from trigeminal neuralgia in that it usually occurs in young adults. Anyone who has seen these patients—and the ineffectual way in which they have been treated—would not be surprised that out of a group of six of my patients (it is a relatively rare condition) four had made suicide attempts because of the pain: one of them after her jaw had been bound together by wire to immobilise the spasm. Re-education is arduous and takes many months but all of the six cases became free from pain.

It is interesting that Alexander in his writings described torticollis and trigeminal neuralgia as two conditions which he had been able to help markedly: and one must remember that in the past the Alexander Principle has often been the last-ditch approach, when everything else had been tried, and the patients were severely distressed and disturbed by the chronicity of their condition: and, of course, sceptical of getting relief from yet one more suggested approach.

One further category which should be mentioned is accident-proneness, whether it be in motor-cars or in the clumsiness of handling objects and one's general movements in everyday life. The well-integrated person is less accident prone. Indeed, it seems obvious that the better balanced person is going to do himself less accidental damage and be less vulnerable to the unexpected incident or fault in his environment.

Local versus General

This chapter has touched on a hotch-potch of medical processes; but then the body itself is a hotch-potch of interconnecting processes and usages.

It is clear that such a conglomeration of bodily processes and uses must be integrated by a Principle—a principle which permits of order and hierarchical structure: a hierarchy in which isolated parts of the body are not permitted to gain dominance over the well-being of the whole on an end-gaining basis. The Alexander Principle says that it is time to understand just how USE can be regulated to produce a stable equilibrium: and it lays down procedures by which the conglomeration of potentially conflicting use-patterns can be integrated into a hierarchical structure. It embodies this hierarchical structure in a newly learned body-construct, through which reflex "stupidity" comes under the influence of the "wisdom" of the brain.

Part of the objection to Alexander's work, in his lifetime, was that some people felt he was claiming to teach people to control their functioning directly with their conscious mind. He never suggested—and it is not suggested here—that we know enough about our physico-chemical processes to attempt to control them by will, except perhaps on a research basis (rats can influence their heart rate, blood pressure, intestine movements and urine formation; human beings likewise can be trained to reduce their blood pressure, by conditioning their autonomic control of blood vessels). Such isolated control is at present more of a curiosity than a clinical tool. It is rather that by establishing a principle of *chosen order* in the body's muscular USE, the autonomic system does not remain so disordered after stress. A balanced resting state of use—in which individual parts do not gain dominance over other parts to which they should be subservient in the total hierarchy—appears to modify the development of persisting autonomic imbalance.

The Alexander Principle applies right across the medical field. When someone gets ill, the sick part of his body is, of course, important; but just as important are the things which he does with the rest of his body in response to the sickness of that part. The patient feels ill because he has, say, an acute sinus infection, but at the same time he will feel ill because he is using himself badly. The use-patterns which he has developed over the years provide a context which both predisposes him to his illness and also diminishes his resilience and capacity to adapt to the stress of the present illness. But all of this tends to be ignored by his doctor. The doctor thinks the patient is ill, say, because of his sinus, and the whole rigmarole of medical investigation is switched on to this. The sinus may indeed be making him feel ill, but a vast amount of the feeling of "illness" is based on bad USE. As soon as it is feasible, the patient's USE needs as much attention as does the specific pathology.

When and if the sinus trouble eventually clears up, the doctor will drop out of the picture, and from his point of view the patient is, for the time being, restored to normal health. The doctor will only have seen him on a few isolated occasions, but the patient has to live with himself always, and he has to live with his persisting habits of wrong USE. His doctor finds "no demonstrable disease": no descriptive disease-label may now be pinned on him. But in spite of this, wrong habits of USE persist and lead the patient to feel continuously fatigued and unwell, to the point when the strain of making any extra effort outside of his everyday life may far outweigh any personal gain or social pleasure.

Investigations which were carried out at the Peckham Health Centre showed that, out of 1,666 normal individuals who were examined, there were 1,505 cases of classifiable disease: but, in addition to this, the investigators found a widespread condition of "de-vitalisation", characterised by an overwhelming sense of fatigue and loss of vitality.

Nowadays it is popular to have general "health screening" o men and women over the age of 50 to exclude serious illness, but, alas, as yet such health screening does not include an analysis of their mis-use. Any investigation into problems of "dis-ease" and "de-vitalisation" must take into account the persistent influence of USE in every reaction and during every moment of life. Any diagnosis which ignores this influence is incomplete. Any plan of treatment which fails to take it into account must leave behind a predisposition to disease and malfunctioning.

Recommended Steps

So I make the following suggestions for the use of the Alexander Principle in medicine:

1. Preventively: during the school years (at least), Doctors, PE instructors, dance teachers and school teachers should be aware of mis-use in the children under their care. Some notice should be made on the child's record and any deterioration noted.
2. Therapeutically: the USE-factor should be included in the various systems which are expected to be examined by doctors. Not merely, as at present, as a cursory analysis of joint mobility, integrity of the reflexes, muscular power and so on; but as a bodily system in its own right, as important as any other bodily system. No medical student should complete his training without being given some knowledge of the the USE-factor.

3. In hospital medicine, there must be a new concept of rest, so that the patient who is lying in bed supposedly resting is not setting up excessive dystonic patterns.
4. Specifically, it must be used in the care of rheumatic, orthopaedic, neurological, psychosomatic and mental disorders.
5. In order to do this, nurses and physiotherapists should be instructed about USE. Not only will it help them in handling their patients, but it will give them personally an additional way of coping with the stresses and strains of hospital life; and heaven knows, there is plenty of stress for nurses in a world in which trival matters are often apt to gain too much importance, a world full of rigid rules and hierarchy, a world in which the personality has to be constantly adapted to suit different patients, and yet a world in which there is lack of scope for initiative.
6. In industry, by an extension of the present work of ergonomic scientists and factory doctors so that they can learn to detect working conditions which encourage mis-use and to observe correctable mis-use in workers.
7. On a larger scale, the preventive implications of the Alexander Principle must be plugged by those interested in health education everywhere. The resources of television, radio and published articles must be used to make these facts more widely known.
8. To make all of the foregoing possible, people must be trained specifically in teaching the Principle, in order to help those who fall into the above categories. The size of the problem should not be allowed to prevent the start being made. Sooner or later, a wide-scale attack on these needs will have to be attempted.

Chapter 7

MENTAL HEALTH

IT HAS BEEN suggested that if a concept could be found in the field of Mental Health which would be as basic as is the concept of MASS in physics, then the whole subject would be revolutionised. USE may well be just that basic concept.

It cannot be emphasised too strongly that it is not ideas which are responsible for neurosis. It is the way in which we react to our ideas with our dystonic use-patterns which constitutes the neurosis. The reason why this has not seemed obvious in the past is that, prior to Alexander's work, there has never been an adequate micro-analysis of dystonic use-patterns. Until his concept of USE was established, no criterion could be adopted which would tell us in what ways a mentally sick person was departing from good USE.

Much behaviour which passes for normal already contains traces of neurotic mis-use. Such pre-neurotic mis-use patterns are usually ignored or not recognised until they have developed to the point of seeming definitely odd. Before the development of such odd behaviour, the mis-used person may have been thought simply to have the "normal abnormalities" which go to make up most people's personality. It is only after improved use-patterns have been developed that it becomes clear just how much the previous mental disorder was based on an inadequate manner of use: and such a manner of use will be seen to return if the neurotic pattern returns.

Much of what I write about mental health is clearly at odds with much that is currently said and done. But there are no cut-and-dried absolute verities in this field. Psychiatrists, behaviour therapists, psychotherapists of every shade and complexion hold views which not only conflict with other groups but with other members of their own groups. I have been a member for many years of the Society for Psychosomatic Research, and the Philosophy of Science Group: anyone who has attended meetings of these august bodies will recall an atmosphere of disputation which is fine for Science but which does not necessarily bode well for the handling of patients.

By adopting an innocent attitude, which says "This is how is seems to me: I may be wrong", I might appear to be attempting to buy off criticism. But the passion to prove and disprove theories in a rigorous scientific manner is not compatible with the immediate clinical handling by one man of a sufficient number of patients. Once I had established—in as rigorous a scientific manner as I could[26, 27]—that a method did exist which could reliably alter USE, it seemed to me more important in my own lifetime to employ that method over the large range of medical and mental problems which were referred to me by other doctors, in the expectation that I would gradually come to an understanding of the possibilities and limitations of the approach. If my account appears simplistic, it conceals an all too willing disposition to take on all comers on my home ground.

Anxiety and Muscle Tension

I spent the greater part of ten years working very closely with Alexander, often just with himself and myself and one patient in the room, and it soon became clear to me that his work could provide a way of detecting and observing "mental" states which no other process could throw up in such tangible form. Behind his theories of the "Primary Control" and a seemingly gymnastic preoccupation with getting people to sit-down and stand-up without upsetting the tension balance of their necks and backs, lay a constant preoccupation with the "chosen" and "unchosen" components of behaviour.

One of the earliest medical articles which I wrote on his work, in 1947, was entitled "Anxiety and Muscle Tension",[28] and it was clear to me then that his approach was very relevant to the handling of neurotic disorders.

The connection between anxiety states and muscle tension is now generally accepted, and the drug firms have been quick to fill the doctor's letter-box with expensive circulars which purport to show just how anxiety can be relieved by using their tension-relaxing drugs. But in 1944 the general psychiatric opinion was that "in anxiety states there is no known structural or chemical variation which accompanies the all too obvious symptoms"[29] and, that "if we consider the long list of psychoneurotics with their hysterias, anxieties, obsessions and states of inexplicable fatigue and depression, there is no characteristic physical accompaniment which can be detected by present methods". It was clear to me at that time, from Alexander's work, that there was indeed a very definite physical accompaniment, i.e., muscular over-activity and mis-use, which accompanied psycho-neurosis.

In the succeeding years it became clearer, by the use of electrical methods of recording muscle tension, just how close the correlation could be between mental states and muscular states. For example, arm tension was found to be connected with hostility: buttock and thigh tension with sexual problems. In many other ways, it became clear that the mentally sick were physically tense. One of the most striking observations at that time was made by Wolff[30] who found that over 90% of headache-sufferers produced their pain, albeit unwittingly, by "marked sustained contraction in the muscles of the neck", and that such muscular contraction was associated with "emotional strain, dissatisfaction, apprehension and anxiety".

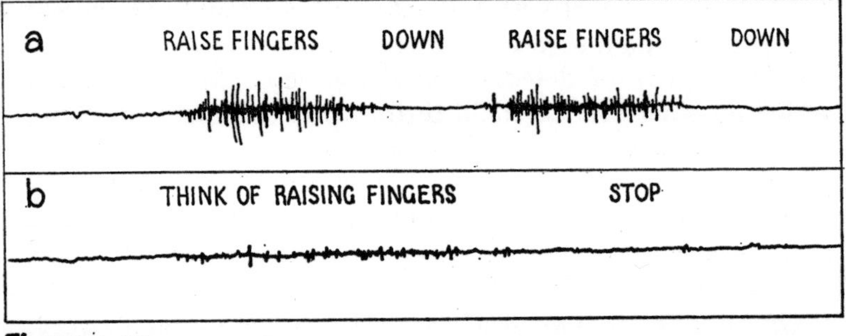

Figure 18

Over the past twenty years, many people have followed up these observations on Mind and Muscle. A recording which I made many years ago illustrates the point (fig. 18). It shows (a) the electrical activity in a person's forearm when an actual movement of the hand is being made, but also (b) the activity which occurs when the hand movement is only being *thought* of. This is an astonishing indication of the link between mind and muscle, and various people of the behaviourist school have suggested that it is impossible even to *think* of an activity without causing small contractions in the muscles which would produce the actual movement in reality. It has been claimed indeed, by some people, that thinking itself is always carried out by such small muscular movements. Small movements of the vocal muscles or of the muscles of the eyes were considered likely vehicles by the early behaviourists. It is tempting to follow up such ideas, but it seems to me more likely that the undoubted over-activity which occurs in certain muscles when most of us are engaged in thinking, is in fact an unnecessary mis-use, and simply gives further indication of how easily most of us will produce unresolved dystonic patterns in response to quite minor ideas.

Gilbert Ryle likewise deprecates, from a philosophical point of view (*Royal Institute of Philosophy Lectures*, vol. 1, 1966–1967), such an equation of mind with muscle, so that mind is even made to include such things as "the tennis-player's wrist and eye movements, the conversationalist's tongue movements and ear-prickings: the typist's finger movements and so on". His paper should be consulted.

But even if muscle is not regarded as a "vehicle" for mind, it is indisputable that tiny muscular use-patterns do frequently occur when we engage in thought (over and above tonic action-currents in muscle). In the words of R. C. Davis: "One has but to observe them on a set of recording instruments to believe that they are by far the most numerous responses of the organism. It is clear that any overt response, vocal utterance, or bodily movement, is surrounded by a wide penumbra of them."[31]

Tension during Communication

The more I worked with Alexander, the clearer it became to me that the tension patterns which arise during movement are quite different from those which are triggered off by the process of communication. The processes of searching and selecting so as to connect-up and communicate with other people are deeply embedded in our character structure and may not be at once apparent as obvious mis-use patterns. Communication does of course take place in a most obvious way by word and by gesture; bodily "mood-signs" are the stock in trade of any actor, and easy to observe. But, in addition to these *obvious* gestures and postures, we find in most people tiny use-patterns which unconsciously, automatically, and whether they like it or not, transmit a certain mood, often quite contrary to their intention of expressing themselves in such a way.

Recent writings on "body language" and "body awareness" have stressed the importance of such non-verbal communication. But many of these forms of USE can only be observed under the conditions of minute analysis which Alexander was the first to employ. One of Alexander's greatest contributions lay in his method of minutely observing tiny dystonic patterns and in his realisation that most of them will only show up *when a patient is asked to learn something new.* Above all, he observed them without introducing tiresome instrumentation and within the context of an acceptable everyday learning situation.

Many of us will be aware of a slight clenching of our hands when we are angry, or a hunching of our shoulders and fixing our chest when we are afraid, or fidgeting movements when we are anxious. But many other tension states occur which are below the level of our consciousness, although contributing a great deal to the general background of our personal awareness. Such patterns may be amongst the most delicate and sensitive means of communication which we have at our disposal, even though they so easily become irrelevant, unchosen, unconscious habits.

Often these tensions include fragments of some role which was significant in the past but which is now irrelevant. They may appear suddenly at quite inappropriate moments, and they may evoke from those around us reactions we do not want. Unless there is knowledge of a basic balanced resting-state of USE, such tensions are liable to build up into attitudes which undermine our interpersonal relationships, without our knowing it.

In many cases the emotional tension may be quite obvious and land us in trouble there and then—for example in the sudden muscular cramps which affect writers, carpenters, musicians, cow-milkers, dentists, golfers (at the top of the back swing), billiard-players (so that they cannot bring the cue forward) and many others, including a recent patient of mine who, when lifting a glass or a tea cup, would get stuck with his cup half-way towards his mouth.

These tension states often occur in a setting in which there is emotional strain. Writer's cramp has been attributed to anger which is unexpressed or denied—anger which leads first to a clenching of the forearm and hand and then to the adoption of increasingly bizarre shoulder and neck postures in an attempt to counteract the arm clenching. In the same way, unwanted sexual feeling may be counteracted by making excessive leg and pelvic tensions.

In most of these dystonic patterns it is not easy to separate a physical reason from a psychological reason. The way in which we construe our surroundings, the way in which we seek to *construct* our surroundings to our taste, is a psychophysical act, in which Mind cannot be separated from Muscle for long.

Mind and Muscle

In spite of his old fashioned stimulus/response approach, Alexander was throughout insistent on psycho-physical relatedness. The relation of Mind to Muscle had usually simply been thought of as "mens sana in corpore sano" (and its corollary "mens insana in corpore insano"). This sort of uncomplicated cheeriness appeared in a popular book by Goldthwaite, entitled *Body Mechanics* (1934).[32] "When the body is used rightly, all of the structures are in such adjustment that there is no particular strain in any part. The physical processes are at their best, the mental functions are performed most easily, and the personality or spirit of the individual possesses its greatest strength." There are reminders here of William James: "Thus the sovereign voluntary path to cheerfulness, if our spontaneous cheerfulness is lost, is to sit up cheerfully, to look round cheerfully, to act and speak as if cheerfulness were already there."

To which one can only say, "Yes, indeed, would that it were so easy." Unfortunately such an approach begs the whole question, which is how such a "sitting-up" is to be obtained. The slumped posture is not the opposite of sitting up straight. There are a myriad ways of slumping and a myriad ways of straightening the various parts of the body. The body is not simply a system of mechanical levers, to be adjusted into different positions like a mechanical crane. It is a subtle organ of expression, in which emotional states modify and are modified by muscular tension states. William James indeed knew all about this when he wrote: "By the sensations that so incessantly pour from the over-tense excited body, the over-tense excited habit of mind is kept up: and the sultry threatening exhausting thunderous inner atmosphere never quite clears away." His sovereign remedy of "sitting up cheerfully" is unlikely to get rid of the "thunderous inner feelings", as long as unresolved dystonic patterns persist.

Nevertheless, one should not forget the role of Courage in the Mind/Muscle relationship. Courage is clearly present in four minute milers, in mountaineers, in riders from Ghent to Aix, and even in the everyday man who exerts will power to take the plunge into a cold bath or the sea. We are familiar with the stories of paralysed and amputated heroes learning to walk, painfully, step by muscular step; much of it admirable and much of it truly heroic, and to be seen daily in rehabilitation departments throughout the country.

But in the everyday world, the "courageous" use of muscle has tended recently to fall into disrepute except as part of the Saturday gladiatorial scene on football pitch or athletic arena. A recoil to the shuffling discotheque has followed: but there has also followed an interest in the less violent use of muscle—a use which will permit and not preclude clarity of thought and emotion. It is with such less violent use of muscle that Alexander was concerned.

Attitude and Emotion

Darwin in his book *The Expression of the Emotions in Man and Animals* (1872)[33] used the term "expressive action" to denote movements, gestures and attitudes from which the existence of an underlying state can be inferred, and he considered that "such movements of expression reveal the thoughts of others more truly than do words which may be falsified". More recently phrases like "non-verbal communication" and "body language" have appeared, and it has become a commonplace that emotional attitudes of, say, fear and aggression are mirrored immediately in muscle; such moods as happiness, excitement and evasion are thought to have their characteristic muscular patterns and postures.

There is nothing very new in this thought. St Augustine wrote in the fifth century, "Hoc autem eos velle ex motu corporis aperiebatur: tanquam verbis naturalibus omnium gentium." (Their intention became apparent through their bodily movement, as it were the natural language of all peoples), *Confessions*, 1.

It is a commonplace that most of us tend to adopt the attitude of those we are with and especially of those we love: even in the cinema we may mimic those we identify with (which may account for vigorous feelings which come on after a James Bond picture). A family posture is often the expression of a basic family mood. Rejection of the family mood may lead to rejection of the family posture—"I can't stand like that, it feels just like the way my mother looks", as one girl with a chronic back pain said to me after her posture had been temporarily corrected. She preferred her pain to her parent's posture.

And indeed to change such engrained postures is bound to have repercussions on interpersonal relationships. Plate 20 shows a civil servant who had severe shoulder pain as a result of an engrained attitude of cringing. It was possible to get him to release the tension by which he was deforming himself and in his new state he was free from pain. Such, however, was his attitude of cringing in front of superiors that he would sooner lose what he had been taught, and give himself a pain, rather than not appear "humble". It was not until he eventually had a flaming row with his boss and "stood up for himself" that he was able to maintain his improved pain-free attitude.

Such deforming attitudes of cringing and evasiveness eventually lead to structural change: the evasive action of turning the gaze away by rotating the head will eventually set up a permanent twist in the neck: and this in its turn makes evasiveness more easy.

The *Attitude Theory of the Emotions*[34] is fine as far as it goes, but its observations are mostly still at a crude behavioural level. The cases I cited above—which could be multiplied almost indefinitely—do indeed say that the body has its language and that we have for too long neglected the important messages which are sent out in that language. But the crude James-Lange approach which says that the attitude *is* the emotion does not do justice to the facts. Almost any emotion can get latched on to almost any habitual muscle tension trick. The named emotions—anger, fear, jealousy, evasion, cowardice, courage and so on—are only a fraction of the multitudes of shades of emotional feeling which go on in all of us. Most of these feelings have no particular names. They are part of the background of our general awareness—a background which is sustained by our particular USE.

And even at this crude macroscopic level of observable "body language", the message may not be an immediate one. *Many habitual postures do not represent an immediate expression of an emotion*, but are rather a *position from which certain actions and emotions can be possible.* The slumped collapsed adolescent is confused at first if he adopts an improved USE, because most of his favoured interpersonal reactions become impossible with this new use. It is only from his old slumped and twisted state that he feels able to communicate with those he likes: his old slumped posture is the position from which he can express certain emotions and from which he cannot express other reactions. His particular posture is not necessarily to be acquainted with any particular emotion, although eventually a state of depression will be only too easily induced by a state of slump.

Many of these postures indeed do not start as an emotional response, but rather from the way we use our bodies in recurrent work situations. The office worker, the conveyor-belt engineer, the lorry driver, the mum-bent-over-baby, the dentist, the pianist, carry out certain occupations for so long that they eventually will hold themselves partially contracted even when they are not involved in the actual pressures and resistances of their jobs. This residual tension may not be conscious but eventually it is maintained most of the time. The summation of their various temporary attitudes eventually finds its expression in a posture—or in a limited repertoire of postures—which come to dominate a person's character. These deforming postures are the result of the gain in strength in certain specific muscles which have never been properly released, until eventually it becomes easier to rest and to move within the familiar deformed structure. In small and at first unobtrusive ways we become enslaved to our past.

The Restructuring of Use

Alexander saw, as many have seen, that what was needed was a better way to help a person to reconstruct his life so that he need not be a victim of his past. But he attempted such a restructuring at the fundamental level of USE, on a general rather than a specific basis; and in a way not attempted by other forms of psychological treatment. No one could claim that drugs or shock treatment restructure a person's USE, except in the crudest sense. The alternatives of individual or group psychotherapy in which a person's ideas and constructs are examined and worked through, is, when available, an advance on crude physical treatments; but the inescapable fact remains that all neurotic people mis-use themselves. I have never yet seen a neurotic person who did not show dystonic patterns. Accordingly, the Alexander Principle asserts that, in the therapy of neurotic disorders, any form of drug treatment or psychotherapy which ignores the USE factor is inadequate.

The Alexander approach reverses the psychotherapeutic approach. The psychotherapist says "you will only get rid of your unwanted behaviour in a satisfactory way when your mental attitudes have been sorted out". The Alexander Principle says "It will be impossible to sort out your mental attitudes in a satisfactory way as long as you persist with that faulty manner of use." Not that I do not applaud the insights of psychotherapy and psycho-analysis: but if we recognise mis-use as a comprehensive attempt to deal with our personal experience of emotional distress, there is no need to give priority to discovering a psychological cause of that mis-use. The priority must be given to learning an improved manner of use, so that from this vantage point, the mis-use patterns can be detected, and resolved when they arise.

Psychotherapy seeks to give us insight into why we have taken up this or that attitude. This, however, can be only one of the aspects of the formation of tension habits. It may well be that a given muscle-tension habit started at some specific time, maybe as the result of some severe trauma, maybe as a casual unimportant accidental trick which was discovered and seemed to fit a given circumstance, at a time when the general use-pattern was already becoming disturbed. But at whatever time and however these tension habits were learned, an adult will have elaborated them (and many other tricks) until they are closely meshed in with all of his behaviour throughout his whole day as part of his habitual use. Release there may be from insight into the situation which led to the setting up of a defensive dystonia, but after a time this insight will have exhausted its possibilities for beneficial change. The inertia, the necessity, the comfort (however slight) of the established way of life will continue to provoke tension habits and to keep them in existence: and they will stay in existence until a thorough re-education of mis-use patterns has taken place. With the best will and the best insight from the best psychiatrists in the world, we can only eventually tackle our tension habits by unlearning them at each actual moment of behavioural reaction and this means the establishment of a CONSCIOUSLY STRUCTURED PATTERN OF USE.

The Direction of Use

It is at this point that many people have been baffled in the past by Alexander. It is here that the Alexander Principle emerged as a truly novel approach.

The novelty arises around what Alexander called "directing" or "ordering". He was asking for something completely novel from himself and also from his pupils: and the trouble was that he did not make very clear to them what it was that he wanted, so that innumerable versions—or none—of this part of his work began to appear.

Let us take the re-educational situation of the Alexander teacher who is giving instruction to a new patient or pupil. The actual re-educational process is described in detail in Chapter 10, but, briefly, two things are involved. The first is a gentle adjustment of the patient—manually—by the teacher, in such a way as to coax misused muscles to a better co-ordination. Secondly, the patient (or pupil, in an educational context) has to play his part in the proceedings *by consciously projecting to himself a sequence of thought which matches closely the occurrences which his teacher is inducing in his musculature.*

Now this is a totally novel way of going on. Behaviour-therapists certainly have their modes of muscle training, with punishments and rewards. Massage techniques, stroking techniques, body contact techniques likewise involve close body contact between therapist and patient. I have surveyed some of these methods in *Modern Trends in Psychosomatic Medicine*,[26] and I made the point there that Alexander's approach was quite different from any of them. He demanded a minutely sensitive attention on the part of the patient to the setting up of a new *ordered structure* in his body: *ordered* in the sense of being consciously projected as a command to the muscles: and *ordered* in the sense of giving sequential attention to the body in a certain 1, 2, 3, 4, 5, etc., order.

To put it another way. A conductor consulted me on account of a pain in his right arm, his baton arm. Not only was his shoulder fixed and painful, but his elbow also—a condition akin to a tennis elbow. After I had started to give him instruction, he asked me to a rehearsal and to subsequent performances. I had already observed that, even when he lay still on my couch, his whole upper back tended to pull over to the right and that his rib cage was rotating more on the right side. It became apparent when he started to conduct that the movements of his right side became completely dominant over the demands of the rest of his body. He reacted to the needs of his baton by losing command of the general co-ordination of his trunk. His head pulled back and sideways, his shoulders hunched, his chest rotated to the right side; in particular, his baton movements were not made simply from point X to point Y, but after reaching point Y would jerk excessively back again towards point X, rather in the manner of a reflex knee jerk.

The conductor was reacting muscularly in the wrong order. Instead of maintaining the central co-ordination of his back as a "core-structure", he was becoming totally involved in the peripheral movements of his arms, so that the structure of his trunk was distorted and his basic balance upset. His resting posture became more and more twisted, and by the end of the concert he was losing touch not only with himself but with his orchestra and choir. His muscles were reacting in the wrong order.

It is perhaps a play on words to speak of "ordering" in two senses—getting the muscles to react in the right order, and giving "orders". For this reason it is perhaps better to talk of giving "directions" to oneself. Even this may—to some people—smack too much of vitalism and mind/body dualism. It does not really matter much. If it is preferred, this activity can be called "projecting a pattern", "employing a new bodily construct", "ordering", or simply remembering to release and relengthen muscle during and after movment in a certain sequence. Whatever the activity is called, the patient can learn an actual formal construction of words which he learns to link on to the new desired USE, until, in time, the new verbal pattern can be "directed", "projected", "thought", in such a way as to bring the body back into the desired resting homeostasis.

Under conditions of stress this becomes the patient's conscious possession. One of my patients, a well known concert pianist who had a hatred of flying, tells the story of flying back to London from Amsterdam, sitting nervously with a paper bag in front of his mouth, prepared for the worst, and of the man in the seat next to him saying, "I don't mind you being sick, but I do wish you would stop saying *head-forward-and-up, back-lengthen-and-widen.*"

Each patient will make his own method of "directing",

which, with or without the help of a teacher, he can learn to link up with the new desired USE-structure. One patient tells me that he phrases the orders in the form of a question to himself, enquiring in sequence of his body whether it corresponds to his desired USE-structure, in the way that a laboratory technician might test the colour of a sample of blood against his haemoglobin standard colours. The "order" presents a standard against which muscle-information can be matched, so that, by feed-back, the "mis-match" signals are eliminated and muscular matching obtained.

Directed Awareness

It is this aspect of the Alexander Principle which has in the past excited the interest of theologians, philosophers, artists, and of course the crankiest of cranky mind-over-matter merchants. It is a source of somewhat wry amusement to deal in one morning, as I once did, with a Nobel prize winner, a famous and dim-witted television beauty, and a young spastic boy, and to be told by all three of them that they approved of what I was doing because I was "speaking their language". (The spastic boy told his mother that he felt he understood me completely.) It does not seem to matter much if the teacher finds himself transported to the wilder shores of meditation, Sufism, Subud, Gurdjieff, Noöspherism, Jungian Yinning and Yanging, Reichian genitalism and so on, provided that the patient can visibly be seen to be linking up a new "direction" with the new improved manner of use. I would not presume to be derogatory of any of these things—their proponents as I have met them are, in the main, highly evolved and sensitive people: but they leave me with the impression that since my Alexander-training is so acceptable to them and so enlarging to their concepts of mental health, then the Alexander Principle of USE may indeed be as fundamental to Psychology as MASS is to Physics. And this impression is confirmed by the acceptability of the principle to a broad range of psychiatrists, psychotherapists, academic psychologists and behaviour therapists.

The Inscape

The poet Gerard Manley Hopkins coined the word "inscape" to describe our personal internal "landscape" and this may appeal more to many people than phrases like "body image" and "body construct". The "inscape" which Alexander sought, as a teacher, to construct for his patients and pupils was made up of a formal sequence of words, which, if they did nothing more, reminded the person who projected them to himself of the Alexander experience which had been built into them so far. But the Alexander "inscape" is designed to do more than this. It specifically sets out to counteract such states as anxiety and depression: and to counteract them not simply by inducing a self-hypnotic mental Nirvana but by influencing the actual muscular structuring of the body.

Plate 21 shows a woman who has been taught just such a new "inscape". Her face, which cannot be shown, was one of typical depression: likewise her posture was collapsed and heavy. Her mental depression found its counterpart in her physical collapse. The second picture shows her after only a few weeks of re-education: the third picture shows her six months later, during which time she had received no further re-education at the hands of a teacher, but had worked on her own as she had been taught. The improvement has clearly been dramatic.

There is no need to go on about the typically dejected and collapsed posture of patients in a depressive illness—in or out of mental hospitals it is clear for all to see. But it is sad that amongst the typical physical functions which are held to be most usually disturbed in depression—sleep, appetite, libido, weight, etc.,—there is no mention of USE. And since USE is ignored, it is not surprising that four out of five people with a depressive illness relapse—and go on relapsing with increasing frequency as they grow older. Certainly anti-depressant drugs and shock therapy may help the immediate impossible situation; but depressive illnesses nowadays differ little from the way they were fifty years ago and—as things are shaping at present—they will look much the same in fifty years' time unless people wake up to the factor of USE. In Hopkins' words (in "The Leaden Echo and the Golden Echo"):

> No, nothing can be done
> To keep at bay
> Ruck and wrinkle, dropping, dying, death's worst,
>   winding sheets
> Tombs and worms and tumbling to decay;
> So be beginning, be beginning to despair.

One influential neuro-psychiatrist in the 1960's, when asked what could be done to avoid depression, replied, "I don't think that the individual can do a tremendous lot to avoid illness, except in palpably suicidal situations like smoking and over-drinking." When asked "Would it count as a great progress in medical theory and practice if you could simply give a man a shot of the right mixtures that would do the whole trick", he replied, "Naturally one would prefer the short and effective aid: experience has taught me that this is so very much more predictably effective." And, in the same interview, he was quoted as saying, "It can be an awful nuisance to have an intelligent patient."

The Alexander Principle is addressed to intelligent people and intelligent patients who are sick of being treated like morons and filled up with drugs. Some at least have begun to realise that a drugged half-life—even if it means that certain long-stay mental patients can be put out of hospital into circulation—is degrading to their dignity and to their potentialities. Such people are casting around for other answers, even though they know that such answers can never be easy.

It will now be clear that the Alexander Principle is not easy. Most people who have had practical experience of it see it as obvious, but never easy. But most people accept with avidity an approach which will give them a chance of long-term mental health, however difficult that approach may be. It is no easy panacea, but it is a major possibility of escape for many—an escape from dependence on doctors and drugs, an escape from the boringness of habitual patterns of thought and behaviour.

The Cause of Mental Disorder

A diagnosis, as Chapter 5 has stressed, is needed both for an explanation and for a prediction. For tackling the present existing situation and for planning a "better future". How can USE be seen as an explanation, a cause, of mental disorder? How can a knowledge of USE be used for planning the future?

First we need to understand what a "cause" is—there are all sorts of causes.

If you throw a stone at a brittle glass window it will break, and under similar circumstances it would always break. But there are more causes than the actual stone. If I hadn't *wished* to throw the stone, the window would still be intact; if it had been a ping-pong ball, the window would still be intact; if the window had been as tough as a car windscreen, it might not have shattered. Three sorts of causes: 1. My wish. 2. The stone. 3. The brittleness of the window.

These are usually called:

1. Final-cause motivation: i.e., my wish for a certain end, or for the consequences-which-flow from that end, e.g., so that I can steal a diamond bracelet from the window.
2. The "efficient (or effector) cause": i.e., the actual moving factor which does it: in this case, the stone.
3. The "dispositional cause": i.e., the circumstances which make it possible—the brittleness of the glass, the absence of policemen. This disposition of the glass is a latent capacity which exhibits itself when the circumstances arise.

Let me take another example. I am frying an egg for breakfast, in a frying-pan, over a gas flame. The final cause (my motivation) is that I want the end (i.e., a fried egg).

The efficient (effector) cause is the actual gas flame: I turned on the gas—the first step: lit it, and now it burns merrily. I would not have turned it on without some sort of decision to do so—the decision, born of my "final cause".

And lastly there is the dispositional cause. An egg is disposed to coagulate when heated, because of its chemical composition.

Which Cause?

Which of these three types of cause are we to blame for a mental disorder? Which of them can we alter?

When people put forward several explanations of something, the explanations are not really *rival* explanations—they are answers to different questions. There are as many sorts of explanations as there are questions: it is simply a matter of priorities. The causes (antecedents) which matter are the one which can be altered and usually the chief causes tend to be the one which one particular speaker thinks *should* be altered.

If, say, a car has hit a pedestrian, there are innumerable antecedents without which the accident could not have occurred. The pedestrian who was hit says it was the driver's bad brakes: the driver says the cause was the pedestrian's carelessness in stepping out into the road without looking. There may be no disagreement about the facts; both may agree that the brakes were bad and that the pedestrian did not look soon enough, and that if either of these things had been otherwise, there would have been no accident.* In the law courts we have to decide who was to blame. In medicine we not only "blame" but we seek to prescribe measures which will prevent it going on or happening again.

This stylised pedestrian/driver situation can be paralleled in disputes about the cause of mental disorder. Disputes among psychiatrists about causes could more realistically be listed as disputes about treatments. The true cause lies deep in the psyche (psychotherapy needed): the true cause lies in the brain chemistry (drugs needed): the true cause lies in conditioned reflexes (behaviour therapy needed).

These three rivals may not be clear-cut and may overlap in their methods at times, but, by and large, they remain distinct. Three different causes are postulated: three different approaches are recommended.

These three different schools of psychiatric thought range themselves around the three "causes" which I have mentioned —"dispositional causes", "final causes", and "efficient causes". The dispositionists search for a chemical or physical predisposition in the brain which may lead to abnormal behaviour: and since at present they only have crude chemical notions of what this might be, they use crude blunderbuss chemicals with which to alter the brain chemistry, and do not mind if they alter other chemistry, which is not faulty, at the same time. Alternatively they may use crude physical methods such as ECT or leucotomy to alter engrained brain dispositions.

The "final-causists" (the various schools of psychotherapy) work to find the decision which was once taken to "throw the stone or fry the egg", a decision which perhaps was once consciously chosen and deemed appropriate, but has now become unchosen and inappropriate. And indeed, if we try to sort out neurotic mis-use in terms of such final cause motivation, we can perhaps find out eventually (over months, years or even decades) what the patient's "game" was which started that particular tension manifestation. But by now the tension trick is so much part of him that it will not be released simply by him

* I am indebted to Anthony Flew (personal communication) for this example.

discovering why he started it. The patient may come to terms with it, but much of it remains embodied in the whole life pattern which he has evolved, and the muscular usages through which he expresses it.

The third cause—the "efficient" cause—has as its champions the schools of behaviour therapists, and right manfully do they battle, in seeking to alter the conditioned reflexes which have become embedded in the life pattern, whatever the original reason for the conditioning. But they are at a disadvantage—an unnecessary disadvantage. They do not know enough about mis-use, and they work with an ethology of *macro*-behaviour—relatively gross observable behaviour—and not at the all-important *micro*-analysis of USE.

The writers'-cramps and phobias which they tackle by aversion-therapy and deconditioning are crude stuff indeed. Their operant conditioning techniques often appear terrifyingly aggressive to anyone used to the subtlety of the Alexander concept of muscular use.

Three types of cause for mental ill-health: final cause, efficient cause, dispositional cause; ideas, muscular behaviour, physical predisposition.

The Alexander Principle—the teaching and learning of which is considered in Chapter 10—takes all three into account. Final cause—the choice—is dealt with by the remaking of the body-image, and by teaching the new faculty of "ordering". The efficient cause—the muscular reaction—is dealt with by a minuteness of muscular analysis which is undreamed of by the behaviourists.

The dispositional cause is what we have referred to in Chapter 4 as the "resting-state", which may be either unbalanced or balanced. This resting state is manifested in the patient's *general use pattern* and it is dealt with over a period of time—by a gradual re-education of the postural attitude and disposition—and not by instantaneous chemical or physical onslaught. Through it, a new disposition, based on a new USE-structure, can be built up. The Alexander re-educational procedure occupies itself with all three "causes" in its practical learning situation.

The Alexander Approach to Mental Health

The Alexander Principle is no cure-all; but it does provide, for someone who is prepared to undertake a considerable discipline, a chance of health where previously there was only despair.

I said at the start of this chapter that much of what I would say would be at odds with the current treatment of mental disorder. The Alexander Principle makes the following points:

1. No psychological diagnosis is complete unless it takes your USE into account.
2. You may get help and insight from psychotherapy, but such insight will not in itself alter your habits of USE. Your unchanged habits of USE will provide a soil on which further mental disorder can grow.
3. Insights which you may get when in a very disturbed state of USE are not necessarily to be trusted.
4. Accordingly, it should be a priority in mental treatment to obtain the best possible resting state of balanced USE before embarking on treatment. The same applies to group therapy.
5. The success of any given treatment (whether by psychotherapy, behaviour therapy, drugs or shock treatment) should be assessed by its effect on the general USE and not just by its success in putting the patient back into circulation. There is as much USE-insanity in the streets as in the wards of our hospitals.

Conclusion

This chapter has taken us to the most fundamental aspect of Alexander's work—its basis in personal direction and choice, and its construction of a basic "inscape" or core-structure from which choice can be possible. It should now be clear that the Alexander Principle is neither predominantly psychological nor predominantly physical but *psycho-physical*, bridging the gap between the analysts and the behaviour therapists, between the prayers and the drugs, between psyche and soma.

The Alexander Principle proposes that if you will acquire a more balanced USE, you will be able to go into new surroundings and accept new experiences (or old experiences which have previously thrown you) without the old degree of strain. Not right away, but gradually you will find new ways of connecting up with things and people, without fear or stress.

Nowhere is this more relevant than in the sphere of Sexual Functioning, and this topic will be pursued in the next chapter.

Chapter 8

THE PSYCHO-MECHANICS OF SEX

SEX, WHEN RIGHT, is for most of us the most pleasant desireable, enjoyable activity in which any of us will have the luck to engage. It takes a lifetime of study and sensitivity to explore its possibilities; it goes devastatingly wrong for all of us at times; it goes miraculously right for most of us, often when we least seem to expect or deserve it; and, like everything else, it is facilitated and it is hindered by the sensitivity of our manner of use. It is exacting and it is mysterious: we can never quite know what will switch the current off or on.

In sexual matters, there is nowadays no firm agreement about what is right and good. Variegated guilt-free sex with many partners, aided if necessary by drugs which are believed to be relatively harmless, is set against long-lasting "normal" sex, in which the same male organ is more often than not in apposition with the same female organ. On the one hand, "magic" is sought from technical variety and drugs: on the other, from an experience which can grow and develop over a lifetime of "normal" intercourse with one partner. Both sides see the other side as wrong: "obscene" on the one hand, "old-fashioned" on the other. Both seem to desire a "good" which is incompatible with the "good" which the other wants.

In a scientific society, no doubt some sociologist-cum-anthropologist-cum-psychologist could eventually produce evidence to show that matched controls—just imagine trying to get sufficient numbers, let alone following them up for at least twenty years—were happier following a variegated sex-life than the old-fashioned "normal" or vice-versa. In the absence of such unobtainable evidence, all that can be suggested are certain immediate biological principles, which will hold good for sexual behaviour whatever the culture.

Such is the power of our early training that most of us are cautious about where our sexual feelings may take us. The young adolescent girl who had been told by her mother that she must never let a man touch her, and thereafter sat rigid in buses in case a man should accidentally come into contact, seems ridiculous to us: but the initial explosion of sex at adolescence makes most people wary of where their sexual reflexes may carry them, and they may soon learn to stifle even the slightest stirrings.

The psycho-analysts have not helped in this matter. Until recently it has been thought undesirable for an analyst to have any physical contact with the patient who is being analysed, and my own re-educational work, which involves almost continuous handling and adjustment and training of the body, has seemed to some analysts to be inviting disastrous transference situations. Yet this does not take place in any harmful sense. Not only is it necessary for patients to be handled in order to learn how to use themselves properly; it is positively beneficial for them to realise that in this situation they can be touched and adjusted without fear and danger: that they are not going to be raped or feel the need to make instant onslaught on whoever happens to be around.

In *Eros denied*[35] Wayland Young has pointed out how it is that "around the thought and act of sex there hangs a confusion and a danger, a tension and a fear which far exceed those hanging over any other normal and useful part of life in our culture". In the 1970's the sexual revolution which Reich preached, has tried hard to get under way, but the revolution has been, in the main, concerned with the *physical* mechanics of sex: sex supermarkets, school instruction, the pill, easier abortion, less reluctance to make love, less social censure, better facilities for sexual voyeurism. But it is one thing to have the means and mechanics for sex: it is quite another to practice an art which can encompass the ugliness of the jerking dog and the beauty of Marcus Perennius. Plastic-doll nudity, on stage and screen, with simulated coition, tells us little or nothing about the all-important psycho-mechanics of sex which are as invisible as our breathing, as subtle as the tiny brush strokes of a Chinese calligrapher.

Sexual Manners of Use

Sexual activity involves the sharpening of all the senses. In it, the human faculties of searching, selecting and interconnecting are employed with an intensity which is scarcely matched in any other sphere of life—at any rate for the ordinary man or woman who do not spend much time in the creative arts. In it, the workaday system of role-playing and payment for work-done no longer has any real validity: payment and acceptance, giving and rewarding, are instantaneous.

The key word is responsiveness—the responsiveness of our own body to touch and movement, the sensed response of the partner: but mainly our responsiveness to our own feelings.

We have seen from previous chapters that muscular responsiveness is a matter of "feed-back", by which our perception of sensation is adjusted and controlled by a "body construct"—by the way in which we "construe" what is happening and "construct" the muscular reaction which we would prefer.

Feed-back can be of two kinds—*negative* feed-back, in which, like a thermostat, mechanisms of balanced sensitivity adjust the temperature so that it doesn't get out of hand: or *positive* feed-back, in which each fresh stimulus adds to the intensity of the muscular reaction which is providing the stimulus, so that it provokes yet more of the stimulus: the atomic explosion is the prototype of positive feed-back, in which a chain reaction of stimulus and response becomes uncontrollable in a matter of milliseconds.

Sexual responsiveness likewise, quite obviously, involves both negative and positive feed-back. By negative feed-back, sexual stirrings are not allowed to go over too soon into an uncontrollable positive feed-back situation. But such is the cultural fear of positive feed-back that even the slightest stirrings of sexual pleasure may come to be stifled, and the opposite poles of sexual deadness and sexual explosion come to constitute an "either-or" situation, which leaves a vast intermediate territory unsensed and unexplored.

Erotism

Our language has a shortage of words which refer to this vast intermediate territory and there are few words in our language for experiencing the erotic pleasure which does not lead to immediate orgasmic discharge. To feel "sexy" or "randy" implies a state of lust which already sounds either naughty or reprehensible. Far be it from me to decry the whole language and lore of the dirty story which we learn from our earliest school-days and which remain, graffiti scribbled on the walls of our cerebral cortex, long after more important and beautiful ideas have disappeared. The young lady of Spain is with me for life, along with snatches of the Prayer Book, reiterated in youth, not heard recently but easily evoked: the night-time prayer against the devil who "like a roaring lion walketh about seeking whom he may devour, which of ourselves we cannot resist".

The erotic words which we do not seem to possess are words for experiences which do not nightly devour us with passion and which we have no particular need to resist. To talk, for example, of "eroticism" already brings in a boudoir tinge of a sensuality, far from the everyday pleasure in our muscles and skin which should be part of our moment-to-moment bodily awareness. "Erotism" seems to me perhaps a better word, if it can be freed from the overtones of guilt which the Freudian phrase "auto-erotism" gave it. Words may not seem all that important, but words are signposts, and we sorely need words for this realm of awareness and bodily pleasure which may have no immediate sexual implication, although it is a part of the sexual spectrum.

Muscular Texture

Erotism is concerned as much with the *texture* of bodily experiences as with its *structure*. When we lose texture, we lose livingness in our bodies: we feel blunted or deadened: we may feel action occurring (action to gain ends), but the satisfaction lies more in the achievement than in the actual doing. The accent is on specific objects, isolated from their background, instead of on the texture of the background.

The Gestalt psychologists, although out of fashion nowadays, encouraged us to think of perception in terms of "figure" and "ground". To take a simple example—a countryman who comes up to London finds the traffic noise deafening, the filthy streets disgusting. The noise and the dirt stand out as "figure" against the groundwork of his expectations. But a Londoner has long since lost the traffic noise in a general background—he does not notice it. Instead he notices perhaps the number of an approaching bus or other "figures" important to himself. A police-car siren may stand out momentarily as a "figure", or perhaps a jet-plane, but even these soon may become background for more immediate figures.

In the control of sexual perception we are dealing very much with such a "figure-ground" situation, in which, by thought and by movement we can turn our attention from one part to another, from one "figure" to another, until by muscular sensitivity and feed-back we gradually build up a generally heightened awareness of bodily texture. We should not be simply concerned with structure, but with the development and recognition of texture.

Texture—as the computers have proved, whether it be in the micro-analysis of a painting or of a fabric—is based on *order*; and it is based on the repetition of order, in a certain progression along certain hierarchies. If hierarchical order is lost, texture is lost. If a picture is painted in the wrong order, not only is integrated texture lost, but paint which is applied too late may run and "weep" because its molecular structure will not marry with the paint already applied.

Texture in the context of sexual perception is built up of a repetitive order which gives an on-going progression from one state to the next. In sexual perception, as in everyday living, the textural quality of our muscular experience is obtained by a correct order of the structural USE of our body, and by a refusal to fix the on-going progression of movement by muscular tension. Texture is destroyed by tension.

Such tension may be a deliberate defence, or it may be unconscious although once deliberate. When sexual feeling begins to invade the texture of our muscular experience, it may not always be welcome. Negative feed-back is usually adequate to regulate and enjoy the more everyday felicities of Erotism (although even these may be blocked by over-tension): but when there is transition to an on-going sexual progression, in which positive feed-back begins to take over, muscle-tension brakes are liable to be clamped right down.

Characteristic of the beginnings of such on-going sexual feelings are sensations which can be described as "floating" or "falling", or "lightness" or even of gluey "heavyness". Many of us are afraid of such sensations. The sensation of being anaesthetised gives something of such a sense of progressive falling into ourselves; and most of us will have experienced certain situations in which, momentarily, all visible means of support disappear and we feel ourselves falling.

It needs a certain courage not to attempt to counteract such feelings of falling or floating. To "let go" into someone's arms may be infinitely reassuring, but it does not mean that, once the weight is off the feet, it becomes possible to let go all the habitual dystonic patterns of our lifetime. Even the most erudite employer of the most erudite sexual skills and approaches will find that both he and his partner are thrown out of gear and out of pleasure by dystonic patterns which seem to appear for no apparent reason and which may effectively kill off feeling and desire and the natural progression towards orgasm: or may modify both the quality and the timing of feelings so that both are unsatisfactory.

It would be a mistake to make it sound as if most people are making heavy weather of their loving most of the time; but it would be rare to find someone who does not occasionally run into trouble, and there are some people for whom love-making is never very satisfactory and in whom the onward "textural" progression of muscular release is constantly replaced by feelings of tension. Instead of erotic feelings of floating and lightness, they will experience feelings of trying hard and effort—indeed there may be cramp and pain—or there may be a deadening and blunting of sensation.

When texture is replaced by tension, sexual responsiveness may come to be equated with relatively violent movement. Many women feel that they must exhibit by the intensity of their body movements and the wildness of their breathing and voice that they are in no way frigid: they may even persuade themselves in so doing that this is what it is all about.

Sexual responsiveness is not a matter of wild movement except in so far as positive feed-back releases the reflexes. It is a matter of the handling of subtle differential feelings of expansion and swelling, of systole and diastole, throughout the whole body. In a word, it is about the balancing of muscular reactions, and it is about the correct "ordering" of muscular reactions, to produce texture.

The Order of Reaction

The ordering of muscular reactions was discussed in the previous chapter; and, just like the orchestral conductor who allowed the demands of his baton to over-ride the needs of the rest of his body, many people, by over-concentration on specifically genital movements and sensations, begin to induce in themselves a general fixation and rigidity which in its turn restricts the natural progression through the many phases of love making. (In passing, it is nice to remember that according to the fourteenth-century Cabalists the total number of angels is 301,655,722. This number is not surprising when we recall that the different phases of love making were considered to have forty-seven different angels, each responsible for each developing phase: and that it was recommended that appropriate, if brief, acknowledgement should be paid to each angel in passing from one phase to the next. This certainly is appropriate to what might be termed "centimetre" sex, as opposed to the ten second sprinters.)

Such "angel-ology"[36] does not imply a lack of spontaneity. But it does see the highest pleasure as involving the mind as well as the perineum, and it has lessons for those who think that sexual functioning can safely be left to our instinctive drives and desires. It is fondly believed by some that love-making has its own natural progressions which will work automatically in a reasonably satisfying manner. Yet doctor's clinics, marriage guidance clinics and the divorce courts are full of examples of failed sexual co-operation, of impotence and frigidity: and even when one partner gets by, the other partner may be putting a brave face on a lack of orgastic satisfaction.

The orgasm has sometimes been called the "white woman's burden", since failure to achieve it is thought to be shameful, carrying with it a stigma of inadequacy and insensitivity. This in its turn leads to a disparaging of the orgasm—it has been likened by some to an involuntary sneeze—and to an almost Victorian acceptance by some women of the uncompleted sexual act. And this in turn leads to a more and more one-sided masculine performance, and to a tendency to shut off—by excessive muscular tension—the early stirrings of erotic feeling which experience has shown usually to be frustrated. And since erotic feelings take a lot of shutting off, there has to be a correspondingly vigilant creation of muscular over-contraction until such contractions become part and parcel of the character structure. Love-making, quite categorically, does not necessarily work properly just by the light of nature; and where that "Nature" is distorted by mis-use, there is liable to be a corresponding lack of lasting satisfaction.

Sexual Perception

It is possible to have the largest library of sexual information that the literate world can produce and comprehensive access to the facilities required, and yet still to misfire. It is the familiar story of the differrence between knowing *how* and knowing *that*. Sexual performance is in no way different from any other skill in its ultimate dependence on practice rather than theory. The stage may be set: the dating, the soft lights; the marriage, the desire; the adjuncts and the adjuvants; the reflexes which carry their rhythms from swamp and forest to the comfort of the civilised bedroom; and yet something is still needed. Civilised man needs to make his sexual experience new not every time, but every moment of every time.

How is each sexual moment to be creatively different? How can two people avoid the silly traps into which their more stupid reflexes will lead them? In the language of the Alexander Principle—how can awareness of USE be adapted into a subtle instrument of sexual communication?

There are helpful parallels to be drawn between our approach to works of art and our engagement in love-making. The thing which distinguishes a work of art from kitsch is that it is something to which we can return again and again, noticing new aspects, revising initial impressions, discovering intricacies and nuances.

Some very obvious parallels suggest themselves—as many people have noted—in the world of music. An immature listener may at first be able to pick out from a Beethoven symphony only certain melodic passages, rhythmic patterns and dominant instruments.*[37] As he grows more experienced, he will begin to notice variations in themes or patterns, including perhaps some inversions, abbreviated versions or transposed sections; he will be able to concentrate on different instruments and follow their development through the piece or through extended passages. He will have sufficient patience to listen to each movement in its sequence, and to notice difference in moods.

As he becomes a more accomplished listener he will be able to discern relationships between disparate sections, and he will have a heightened freedom which allows him to focus now on one, now on another aspect of rhythm, melody, harmony and instrumentation, and perhaps to focus on several of these

* This example is prompted by Howard Gardner in "Figure and Ground in Aesthetic Perception", *British Journal of Aesthetics*, Winter 1972, although he does not draw a parallel with sexual experience.

aspects at the same time. Eventually he comes to know it so well that he can re-create it in his mind, criticise certain performances of it, anticipate its implications and possibilities. In spite of such sophistication, he may at times employ only a primitive perception, and focus merely on the melody and the rhythm, ignoring other aspects; but unlike less developed listeners, he has the option of returning to a more differentiated and articulated apprehension of the piece, if he desires.

In discussing such a sensitive musical appreciator, I don't think that I need draw heavy-handed parallels with sex-appreciation, except perhaps to point out that, in sexual activity, one is more in the position of the orchestral player, producing as well as listening to the sound, than of a member of an audience: and that for many of us, this will be about the most creative and artistic act in which we will engage.

The Background to Sexual Activity

Three parts of the body, besides the actual sex organs, need to be thought about during love-making, since all three areas may produce unhelpful tension. All three have to be thought about in different ways, since all three can inhibit sexual flow. The three parts are the muscles of the head and neck; the muscles of the chest and abdomen as they affect breathing; and the muscles of the lower back, pelvis and thighs as they affect genital movement. (The arms and legs obviously can also become tense and mis-used in sexual activity; but, by and large, they will behave fairly well so long as the pelvis is free, and so long as the shoulder, hip and knee joints will actually move.

We have considered to some extent the tensions which arise in the head and neck region (Alexander's "Primary Control"). It may be helpful to consider the dystonic patterns which can arise around the activity of breathing. Not only during sexual activity but in many other ways, breathing is inseparable from the handling of the emotional life of each one of us.

The connection between sexual feeling and breathing has been frequently stressed before. Ancient Yoga disciplines prescribed complicated—and to all intents impossible—regimes of taking in air through one nostril down to the genital area where it was to be circulated around before being brought up to be expelled through the other nostril. Recently breathing techniques have again come into prominence with the work of

Wilhelm Reich. I first encountered Reich's ideas in the US in 1949, and was impressed then by his account of "muscle-armouring"—states of tension which prevent the full experience of sexuality.

Reich's account of the stages of the orgasm was masterly, but he sadly mis-fired over his concept of breathing. Lacking an adequate concept of USE, he laid down a number of muscular and breathing techniques which, after the first novelty wears off, are apt to leave basic dystonic patterns scarcely altered. The violence of his breathing and pelvic movements—although perhaps giving brief emotional release to some really frozen-up and unresponsive people—are too disconnected from the every-day living of these people to be of basic usefulness. The subtlety of Reich's written work does not seem to be matched by a corresponding subtlety in practice.

Indeed, similar breathing techniques are to be found even in the folk-lore of middle-class England. One crude way of inducing sexual feeling is to breathe out several times deeply from the upper chest to the genital area whilst at the same time slightly contracting the thighs and buttocks, and something like this was described by Reich. I am reliably informed by a splendid headmistress that when she was young, in the girls' dorm at her particular school, they used to say "Come on, girls, let's URGE", and this was accomplished by breathing out and contracting somewhat in the manner described. This preceded Reich by at least twenty years. I would back Roedean against Reich any day of the week, and no doubt the folk-lore of many countries contains similar and such-like advice to young maidens and men.

I have no such esoteric advice to give, but simply the advice that one must learn not to fix the breathing through excessive or wrongly-distributed muscle-tension: and that if one is emotionally disturbed either by flaws in the sexual flow or other matters, it is always helpful to try to introduce order into the sequence of breathing. It is important not to *start* to breathe in by a movement which raises the upper chest and breast-bone, although this region will normally be raised a little at the *end* of breathing in; and it is helpful to learn to release tension in the shoulders and upper chest as you begin to *breathe out*—this is the only point in the breathing-cycle at which upper-chest tensions can be released without upsetting the cycle. And the more that breathing can be thought of as an activity of the middle of the back, and of the back and sides of the abdomen, the less the opportunity for harmful tension.

How to Lie Down

Sex-manuals are filled with innumerable varieties of coital positions: sitting, standing, lying, kneeling, this way and that. But, by and large, most love-making involves at least one partner in lying down: and usually one or the other's legs are going to be relatively straightened out at the hip joint.

It is interesting that, in the past, when people have talked about the evolution of the "upright" posture, they are usually thinking about standing or sitting upright. But few animals except man can *lie* with their legs straightened-out from the pelvis. We do not automatically know how to lie down properly with our legs out straight, any more than we know how to stand and sit upright properly. If there is to be an adequate and subtle adjustment of the pelvis and its muscles during coitus, more needs to be known about a balanced resting position of the pelvis, when lying down.

A few points may help self-analysis.

Lie down on the back (see fig. 32) with the elbows out to the side and the knees pointing upwards towards the ceiling. In this position there should be no arch in the lower back. It should be impossible to insert a hand between the lower back and the surface beneath. If in fact the back is arched in this position, it will be due to two things: the front of the chest at its lowest point is being pushed forward too much and the whole chest cage needs to lie much flatter against the supporting surface, with the shoulder-blades widening apart; or, the second, and more common reason, is that the pelvis is arching forwards towards the thighs. If the fingers are placed about one and a half inches from the navel on either side and then run down towards the pelvis, they will strike a jutting-out piece of the pelvis on each side. These "spines" of the pelvis, when the back is arched, will be too close to the thighs. The buttocks need to be dropped slightly down and away from the middle of the back—this may be quite a big adjustment if it is very arched at rest—and in so doing the pelvic spines will drop slightly towards the abdominal cavity. If the fingers are kept on the spines as this movement is made, there will be an impression that the spines are slightly separating away from each other. The movement I have been describing—as will be apparent—is very much concerned in the approximation of genital surfaces. During coitus such a movement by both men and women when accompanied by a slight buttock contraction will constitute the inwards sexual connection: the slight releasing of tension the outward movement.

The composite movement of moving the pelvis forward, which involves a slight flattening of the back, will also correspond to a widening across the back of the chest which takes place when breath is taken in. The slight tilting of the pelvis back as the genitals are separated is accompanied by breathing out and a slight relaxation of vaginal musculature or perineal musculature in the man.

The correction of an unduly arched pelvis and back is of course a much larger adjustment than the minuscule "plateau" movements of coitus. But it will be clear that, given a range of movement from A to B, the resting position should be about half way between A and B. If—as in the case of the arched back—the resting position is much closer to A than to B, then the potentiality for muscular relaxation and contraction, lengthening and shortening, is markedly limited. And if certain muscles around the pelvis are held almost permanently over-contracted, the scope for progressional movement is small. The muscles which are usually found to be over-contracted are the buttock muscles themselves, plus the muscles of the front and inside of the upper thigh.

When lying down as I have described, with the fingers on the spines of the pelvis, one side may be found to be higher, i.e., closer to the ceiling than the other: or one spine may be raised up closer to the chest on one side. If this is the case, it is likely that the weight is being distributed more along one side of the back than the other. It should be relatively easy to distribute the weight equally, and this will predispose to a better release of tension in the buttocks and back.

Muscular Blocks and Interferences

Sexual activity, like breathing, is not "about" anything. It can of course be said that it is about producing children or pleasure one with another, just as breathing can be used for producing speech. But the basic processes of sex and breathing are not about anything. A process is going on and a process is something which you must let happen—not interfere with. There are no rules for the right way to engage in these processes: only, perhaps, rules for what not to do. In our moment-to-moment living, we too often make this mistake of *doing* some fixed thing, instead of engaging in some process: we *Read, copulate, eat, speak,* instead of engaging in a *process* of reading, a *process* of copulating, a *process* of eating, a *process* of speaking.

As human beings, we are so constructed that we work best when we concern ourselves with process. When we are concerned with *ends* rather than *means*, our bodies don't function as well. The human organism is built for process-operation, not for end-gaining.

As we engage in a process, something unexpected may happen and if we are not end-gaining, we can notice it and adjust ourselves to it. The psycho-mechanics of sex must learn not to interfere with processes and rhythms which arise and develop spontaneously: if there are to be new perceptions, it must learn to regulate without interference.

Interference takes place at two levels. At the gross mechanical level, postures which have been developed over many years will limit mobility either because the surrounding joints have become stiff with mis-use, or because sexual leverages produce pain and cramp in already over-tense muscles. It is not surprising that mis-used man has in recent years embarked on a variety of sexual practices which need involve only a bare minimum of mechanical mobility. When the lower back, pelvis and limbs can move freely, there is surely enough potential sexual experience here to last several lifetimes.

At a more subtle level, muscular tension patterns of great complexity interfere with the quality of sensation which is fed back to the brain from the on-going processes of muscular movement and contraction. The mechanisms of balanced sensitivity (which have already been touched on in the previous chapter) are liable to become blocked by tension, and any attempt to unblock a feeling of deadness by specific action or movement is liable to fail. Under such circumstances the capacity to give orders and directions to the main core-structure, i.e., the head, neck, back and breathing, is the quickest way to unblock a purely local tension. We have seen that a writer's cramp cannot be adequately treated by concentrating on releasing the muscles of the wrist and hand, but there must first be attention to the co-ordination of the neck and shoulder. In the same way pelvic tensions—once they have become maldistributed—require a release of tension first in the neck and shoulders and then an improved use of the middle of the back and thorax.

Arthur K. was a pianist of 52 who had been married for three years but had never managed to consummate his marriage. He had previously been receiving psychotherapy and he told me that his only experience of orgasm had been in his late teens when he had been severely reproved for masturbating. He assured me that since that time there had been no sexual outlet, and that his psycho-therapist blamed his deadness on his unsatisfactory early experience.

He had been referred to me primarily for treatment of a painful arm and shoulder and it was several weeks before he admitted to his sexual problem. I had already begun to tackle with him the excessive and mal-distributed tension patterns around his lower back and legs and he was gradually learning to release some of the tension. He progressed to the point when I was able to show him a co-ordination of his lower back and pelvis in which his use became normally balanced. At this point he blurted out "I can't do that; it makes me think of the sexual act."

We persevered to familiarise him with the new co-ordination in the easy safety of my training conditions, and he arranged to see me the following week. When he returned, he told me an extraordinary story of returning home and seeing a nude photograph in one of the colour magazines which aroused him greatly to the point when he was able thereupon to consummate his marriage—history does not relate what his wife thought about it—and thereafter he had been able to repeat the experience.

This strange story illustrates the fact that we may know quite well what the factors were which led us in the past to set up a muscular defensive block: but insight into the origin will not necessarily untie muscular blocks which over a period of years have become incorporated into our whole basic manner of use. In this case it was only after the blocks had been undone and the patient felt reassured and at ease in the unblocked state, that he was able to respond without blocking when away from me.

Such stories which arise again and again in the re-educational situation, usually illustrate the fear which people have of unusual feelings. Most muscular blocks and defences were chosen, at the time, for what seemed to be a good reason—usually to avoid pain or rebuke (the so-called "traumatic avoidance response"); rather like the child who learns to call fire a "No-No", after being scared from it by being burnt and by his parents shouting. The on-going sexual sensation of free movement and flowing can also become a "No-No".

Such a patient has to learn to disregard his usual body construct: he has to learn no longer to construe his feelings as a "No-No". He can learn through the Alexander Principle to occupy his thinking with the projection of a new body-construct, which in its turn will feed back new sensations to his brain—sensations which will have no fear-connotation.

Sexual Disorders

The late Joan Malleson wrote to me about the relevance which she saw in such an approach to such sexual tensions as vaginismus: this is another of the muscular blocks which can come to lead a life of its own, outside of its place in the general muscular hierarchy. Similarly, a pre-existent muscular dystonia can lead to such disorders as premature ejaculation and can be helped by a general re-ordering of the body-construct. Such "re-ordering" is relevant not only to the actual timing and performance of a sexual act, but to the role which habit and compulsion play in the desire for sexual outlet, and in the mental states of preparations and "guilty longing" which occupy many people's thoughts to a greater or lesser degree. (A recent survey indicated that most young adults think about sex once every 15 minutes.)

But many people never get to the point of even attempting the sexual act. In our present "free society", shyness and loneliness still abound. Much has been written about the tragedy of being "taken unprepared", but there is the even greater tragedy of being prepared but never allowing oneself to be taken. Many people seem to spend their time putting their money into the telephone box, going through all the work of dialling the number they want but at the critical moment, when they have only to "press Button A" to get in contact, they press Button B to get their money back. And even when they can hear "their own true love" speaking at the other end, urging them to press Button A, their nerve fails and they press Button B again.

However free the social opportunities may seem, shyness and loneliness continue to be induced by muscular blocks and defences which prevent the making and taking of opportunity.

Freedom-in-thought

Alexander was dead before what the 1970's called "The Permissive Society" was upon us. But his views on Freedom were often expressed: that *external* freedom to carry out certain actions (freedom-*of*-thought-and-action) is far less important than our own personal freedom *in* thought and action. He rightly saw that, even when we have the most perfect conditions of environment, company, and cash, our happiness is still determined by our capacity to think newly and freely; and that such a capacity is limited by our manner of use.

It is a great improvement that young (and old) people can now find easier sexual outlets without guilt, and without the feeling that someone will disapprove. The sexual drive is of basic strength: the human body is prodigal of sexual functioning. Not only are millions of spermatozoa made available for a task which one alone can complete, but some degree of sexual pleasure is available even to a severely sick person. The brain is claimed to be the last region of the body to die, but the gonads must run it a pretty close second. The sick man, the evil man, the old man, do not easily lose their sexual pleasure.

I remember, when working as a house surgeon to the late Kenneth Walker, being summoned out of bed to the private ward in the middle of the night by an elderly clergyman whose prostate had been removed and who found to his dismay that he could not obtain an erection. He was not himself, of course, and a few weeks later must have blushed at the thought. He was pacified by my reassurance, but it does indicate how persistent is the sexual urge, and why moral codes and punishments have seemed necessary in the past to curb such a persistent urge: and why the individual may feel the social need to adopt blocking tactics.

Innumerable treatises have been written on moral philosophy, on free-will and choice. Innumerable theologies have presented views on the good life; about heavens and abominations and the wrath to come. You would think that out of all this lot, somewhere someone would be able to give a young (or old) person really useful advice about whether or not, and when, to indulge in sexual activity. Ninety-five per cent of us masturbate or have masturbated at some time: a considerable number will be wondering in the future whether or not to masturbate. Dry textbooks of moral philosophy would surely become best-sellers if they genuinely considered, say, this one sexual dilemma.

But most of the treatises on moral philosophy are about such things as why I decide to cross the road to go to a tobacconist's shop, or what it means to call a strawberry a "good strawberry". They do not tell a young man, who feels guilt after masturbating, much about this fascinating stimulus-response situation in which free-will and determinism are on opposite sides of the pitch.

Perhaps I have an exaggerated view of its importance. The English public school system—the one I know—insisted on little adolescent boys jumping out of bed at 7.00 a.m., and, at my particular establishment, pyjamas had to be removed, and there was a quick run down the stairs to an icy cold shower where—observed by jeering older boys—a jet of cold water played over the anatomy. Now it was an observable fact that most of the small adolescents were in a state of modified erection: modified that is to say by the icy dash. One boy indeed kept his father's large hunter watch by his bed, and applied it to his genitals in the hope of taking the impetus out of his tumescence before the embarrassing run was made. Many of the boys—such of them as one talked to—were plagued by guilt. Others made a thorough Portnoy of their situation; and one of them reckoned to achieve two orgasms—watched by an admiring group of friends—between the ringing of the bell for prayers and turning up spick and span in hall two minutes later for the benediction.

For what it is worth—and as the sympathetic recipient of countless "confessions" from worried patients—I have never seen anyone whose sexual apparatus was harmed in the process, nor do massive masturbators seem eventually to fail more than less frequent performers to achieve successful coitus: it is just that a few of them may lose the exploratory drive which the demand of an unsatisfied sexual urge would give them. And most of them are perpetually worried by the problem of "shall I, shan't I".

The only observation I can make about the morals of whether to have sex or not with or without a sexual partner, and when, is the phrase which I culled from Professor Sparshott's excellent monograph *An Enquiry into Goodness*.[38] He puts forward the view that "To say that a thing is good is to say that it is such as to satisfy the desires and needs of the person or persons concerned". It takes Professor Sparshott a whole book to sort out the implications of this one sentence, but the advantage is that it does refer to other people as well as to oneself and it can also include oneself not only as one is today, here and now, but oneself in a month's time, a year's time, five years' time. Oneself at a later date is very much one of the persons whose needs have to be thought about—obviously oneself in five years' time with syphilis or a genital stricture or a fatherless child has to be considered, as also has oneself with a "frozen" insensitive sexual orientation due to over-caution and timidity. As with most things though—unfortunately—we don't know till we have tried; but at any rate, one can try to avoid making the same mistakes too often and one can try to avoid being frightened off by one or two bad experiences with unsuitable people. One can avoid, say, labelling oneself homosexual after one or two disastrous fumblings with the opposite sex or after one or two enjoyable experiences with the same sex. And, perhaps, one can learn, from a knowledge of the Alexander Principle, that if the USE of certain partners is muddled and disorganised, they may be pleasant companions over a short period, but the unpleasant and difficult habits which are already observable in their USE, will make life miserable for both partners over a long period. Alternatively, on the obverse side of this particular coin, we may perhaps find it easier to salvage something out of a relationship which still has much to commend it, *if we can learn to attend to our own mis-use patterns*: or if this does not help, a more stable manner-of-use may give the courage to make, or take, the break-up.

Chapter 9

PERSONAL GROWTH

THE TWENTIETH CENTURY has seen an increasing sophistication into the possibilities and perils of up-bringing. The Freudian and Kleinian emphasis on the importance of the child's early years has led some parents to instant guilt, since it appears that, no matter what we do, even our everyday encounters with our children can lead to far-reaching consequences. I knew one otherwise intelligent philosopher who put himself, his wife and his three children all under psychoanalysis—the oldest child was six—and such was the intensity of his belief in it that, when one child fell from a tree and cut himself badly, it was a toss up whether they would go first to a casualty department (to have the cut sewn up) or to the analyst (to have the right slant put on it all).

Latterly Winnicott, in a revulsion from such instant guilt, was at pains to emphasise that "mother knows best", and that parents should follow their natural expertise. One of my patients expressed this view nicely when she was writing to me about the similarity of my handling of a patient and the parent's handling of a child.

> Years ago my husband took a film of me putting our second son aged five months into a pram. When I saw the film afterwards I was struck by the obviously deft skilled way I handled the child. I was staggered because I had never thought of myself in this way nor had I realised that a mother's relationship with the child bore a relationship to that of any craftsman to his material. It is fascinating to watch the positive *going with* his material that the carpenter or potter etc., develop through constant contact with it. Mothers and their babies can achieve this: and I in my capacity as a patient with you can be quite content to become absorbed in my body as a material that has a particular grain or plasticity in it, if one can eliminate the need to *react*. In my case this is the key to the success of it. In everyday life I am a powerful reactor and in the sessions I am content to become a material, reacting only to the teacher.

This "natural dexterity" line contains a certain amount of truth, but just as with love-making, natural *dexterity* (along with the other wisdoms of the body) becomes only too often a natural *stupidity*, when it is based on end-gaining. It may be that some mothers do have an inborn dexterity in handling the young infant but many other factors soon come in. The sad fact is that if a mother is herself an end-gainer and the possessor of faulty tension patterns, she will very soon begin to influence her child in the wrong direction.

From the moment of birth the helpless child is dependent on the handling and the ideas of its mother. It is picked up jerkily or smoothly, crossly or kindly: its head and back are supported carefully or ignorantly. It lies face down or face up, according to fashion. It is allowed to yell or it is picked up on demand. It connects with the mother, on breast or bottle, and as it suckles, it likes to gaze long and deep into the mother's eyes, with a unified visual connection which it may never know again. But in the main, its connection is kinaesthetic, through muscles and movement, and it is quick to pick up feelings of tension, timidity or rejection from the bodily rather than the visual contact: and especially from the mother's hands, since another person's hands are a most powerful stimulus towards good or bad USE.

As the nervous system develops, the stage of sitting up is reached. One of Alexander's earliest teachers, Alma Frank, carried out a painstaking study of how children first sit up, and she showed that if children are made to sit up before their nervous systems have adequately matured, they loll about, with the beginnings of mis-use, and their backs develop a sideways curvature. She took some beautiful pictures of children who were left to adopt the sitting-up position on their own, at their own chosen time. She showed that if they were left alone, they would adopt a balanced upright position of the back, with the head in the position of Alexander's "Primary Control".

The child should be left to initiate movements in its own time. To pull a child up by the arms too soon is to ask for mis-use. By the age of twelve months, over 90% of children will have developed a sideways curvature in their backs. The urge to place a child on a pot and leave it sitting there hopefully should be restrained, until such time as it can support its own back properly without lolling and collapsing. Likewise, the various landmarks of standing and taking the first steps to walk should not be hurried: the child should not be "stood-up" or encouraged to walk until its *own* balanced USE enables it to do so.

The Standing Child

There are few more beautiful sights than the well-used child standing with legs slightly flexed in the Alexander balance and with the vertebral column well back, counter-balanced by the head. But by the age of 2½ or 3 years, things are already beginning to go sadly wrong.

The child at this age will have adopted many of the tempos and tensions of the parents. The family mood—or the mood of one dominant parent—will already be inducing its associated posture in the child. This process will continue with us for all our lives, since, if we are to share the constructions which people we admire put on things, we are eventually forced to share something of their manner of USE—a "posture-swapping", in which they may also adopt something of our posture. We imitate the attitudes of those we admire in order to make contact easier: it is through USE that we construe our surroundings and since a major part of our connecting-up with other people consists of an attempt to share the construction which they put on things, we have to adapt our USE to theirs. In a situation in which we are dominant, they will adapt their own USE to ours: "posture-swapping" is rarely fifty-fifty; it tends to favour the dominant person.

The effect on the child of this projective posture-swapping will begin to be a personally idosyncratic mixture of tensions and predispositions of structures and potential attitudes, an amalgam of nature and not-so-much nurture as selective-preference on the part of the *child* as he brings up his parents as best he can!

Schooling

The child is not long under the sole influence (good or bad) of its parents, nor are they for long the sole objects of the child's influence. Much has been written on the stages of physical and psychological growth from the primary schooling stage upwards: too much to summarise in such a short chapter as this by someone as medically orientated as myself. But even to a relative outsider to the educational scene it is obvious that much more is known nowadays about the various stages of personal growth during the school years. Piaget's views on cognitive development have led to wide-spread curriculum changes. Psychologists in general are providing much better ways of assessing the individual child's possibilities and hindrances. Gesell has detailed the stages through which a child's personality will mature, and Tanner[39] has put a welcome emphasis on the widely different speeds at which children grow up and mature.

But, behind all these facts and statistics, the fact remains that each one of us is an individual, and we each of us have our own unique way of seeing things. It might at first seem an impossible task to generalise about personal growth, since every waking moment of every waking child's life is uniquely its own: it has its own unique parents, its own unique home: its own unique reactions to school and learning: its own unique physical possibilities and barriers: and, eventually, its own unique way of connecting-up with things and people.

The story of personal growth has indeed progressed far from the stimulus/response psychology of the behaviourists and the hell-fire of the churchmen. But, *pace* Freud, Gesell, Piaget, Winnicott *et alia*, it has had little to say so far about USE. And since USE is the compendious Gladstone-bag within which the Freudian Unconscious, the maturation processes of Gesell, the cognitive development of Piaget, and the "natural" behaviour of Winnicott, must lie; and since USE is observable—and alterable—by both the child and his teacher or parent (once they know how) it must be given as high a priority in education as in medicine: higher, indeed, since the problem is one of preventive medicine in the widest sense.

Use Analysis

What sort of things should we be on the look out for when we look at the USE of a child or adolescent? It will be clear from the previous chapters that many of the subtle forms of mis-use are not at all obvious on a superficial examination. Nevertheless, repeated attitudes—mental and physical—soon begin to leave an obvious mark on a person. So let us as systematically as possible consider some of the most common and obvious mis-uses which can be detected by almost anyone if they look.

The Head and Neck

Let us take a schematic head and look at it sideways on (fig. 19a). Let us take the 7 neck vertebrae which connect it to the chest (fig. 19b). Now let us look again at the collection of sideways X-rays of the head and neck (plate 18). The most common mis-use involves a pull of the head back and a drop forward of the neck (plate 18a, b, c, d and fig. 19c) but there are many variations. Sometimes there is an over-straightening of the neck (plate 18e, and fig. 19d).

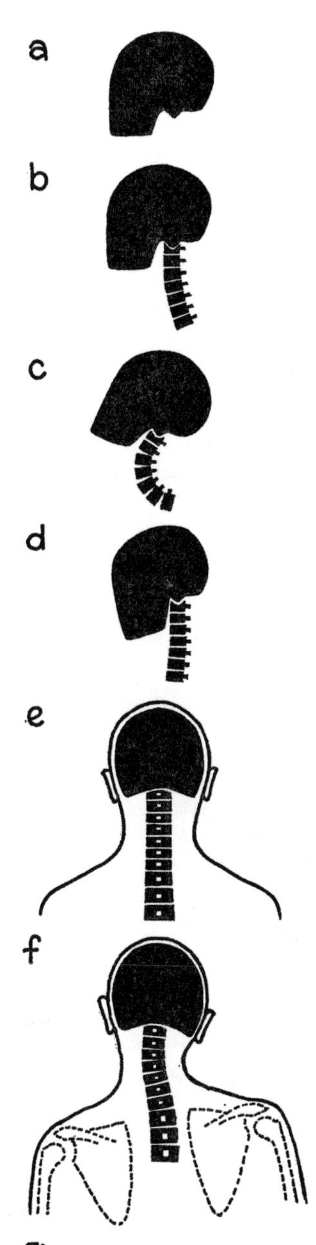

Figure 19

Let us look at a head and neck from the *back* and add a few more vertebrae—those in the "hump" at the upper part of the chest at the back (fig. 19e).

Many people will be able to see in a mirror that they have developed a habit of slightly pulling one ear down towards the shoulder (fig. 19f), as shown by the level of the lobe of the ear. If this occurs it leads to a compensatory twist in the neck, usually in the lower part of the neck and the "hump".

It happens to be a fact that this is a difficult place to X-ray, at least in routine X-rays of the neck and chest: either the neck is X-rayed or the chest is X-rayed—the junction area tends to be ignored. Quite small twists in this area (a cervico-dorsal scoliosis) tend to be ignored, but they indicate quite considerable upsets in the muscle balance in the neck (X-ray plate 22).

Look at the line of muscles as they come out of the neck towards the shoulder (figure 20 and plate 19b).

The line will probably be lower on one side than the other, not just because being right or left-handed has made the muscle bigger, but because of the structural mis-use—the scoliosis.

With this imbalance, there will probably be more tension on one side of the neck at the back than on the other side—a fruitful source of headache and neck tension pain: or there will be more tension in one sterno-mastoid muscle, in front, than on the other side. The sterno-mastoid is the thin muscle which runs from the mastoid bone to the top of the breast bone—it will be felt to contract if the chin is pulled down on to the throat, and if the fingers are placed just above the inner ends of the collar bones. Needless to say many other muscles on the sides of the neck and below the chin will contract if the chin is pulled down like this.

Usually, if the head pulls to one side, it will also rotate slightly, so that someone looking from the back would see more of the jaw on one side than the other. An established rotation or

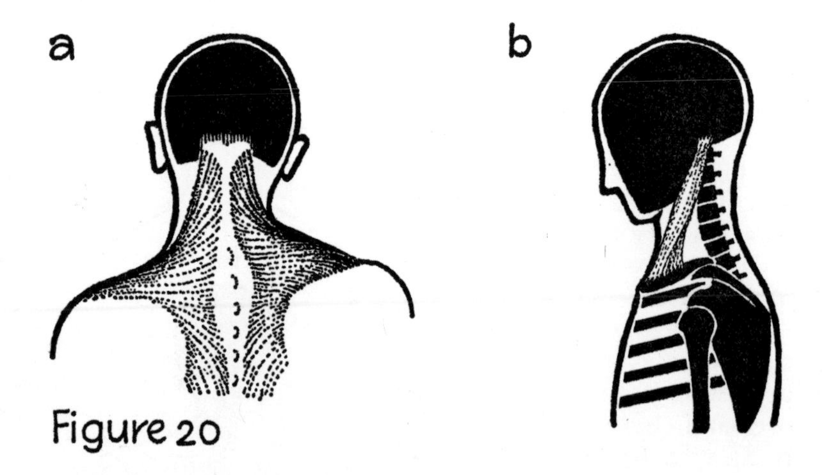

Figure 20

sideways contraction of the head will lead to an asymmetry in the face and perhaps to slight occlusion troubles, since the jaw will be opening sideways. And to that most distressing symptom, the clicking jaw.

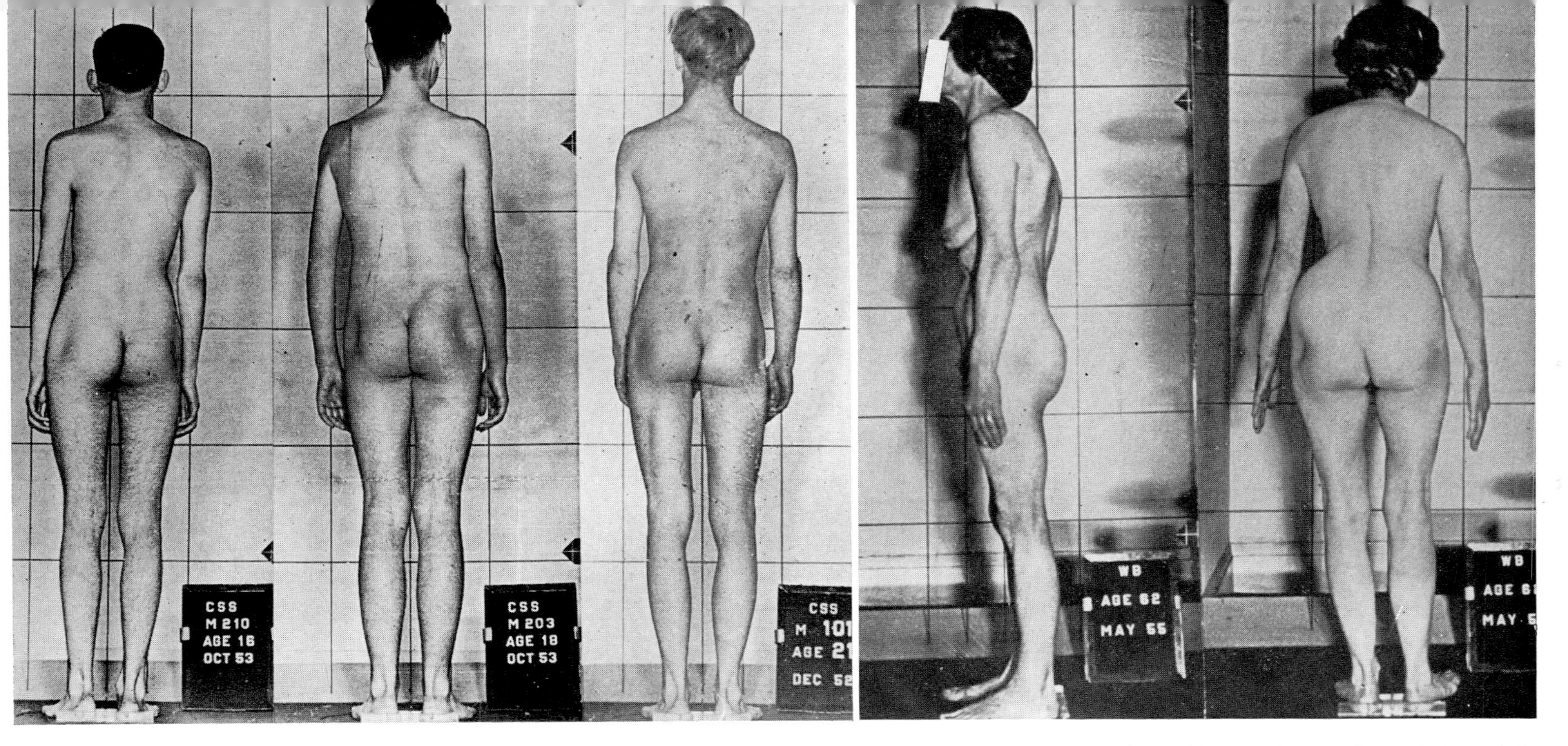

19a Three students with chest displaced to side. Note middle student, right shoulder back.

19b Elderly woman. Chest displaced to right, pelvis to left. Note neck collapse and tense sterno-mastoid muscle.

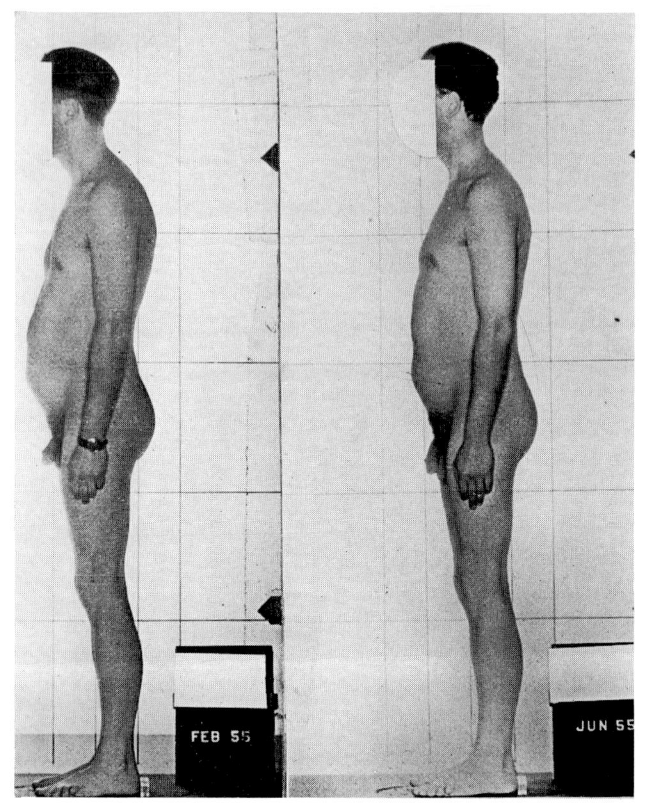

20 Improvement in emotional state, paralleled with improvement in attitude.

21 Collapse and depression. Loss of depression with loss of collapse.

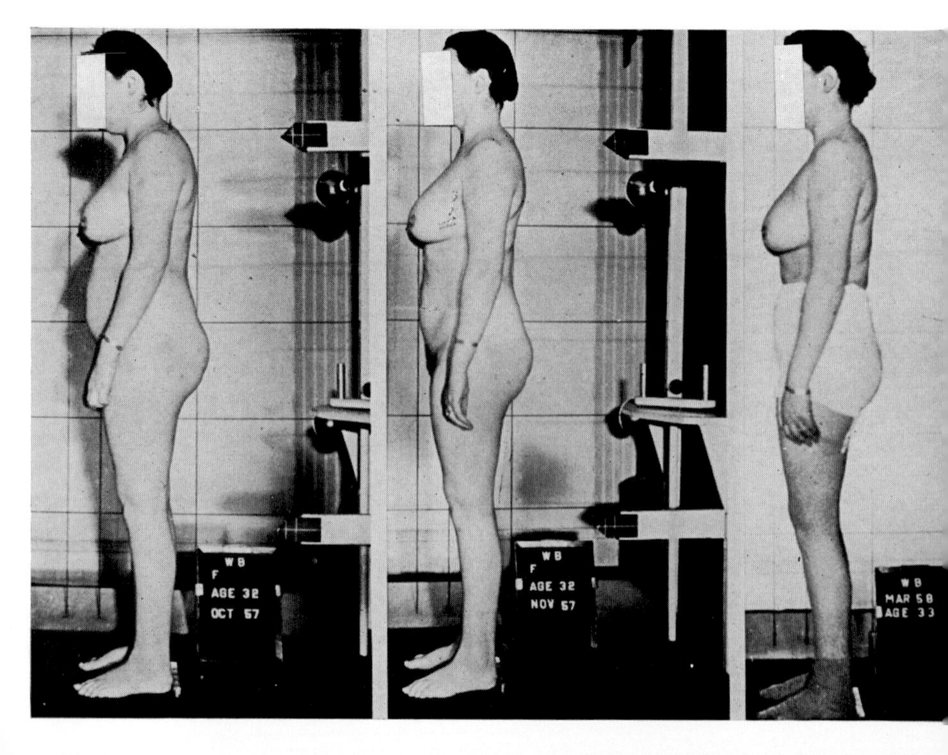

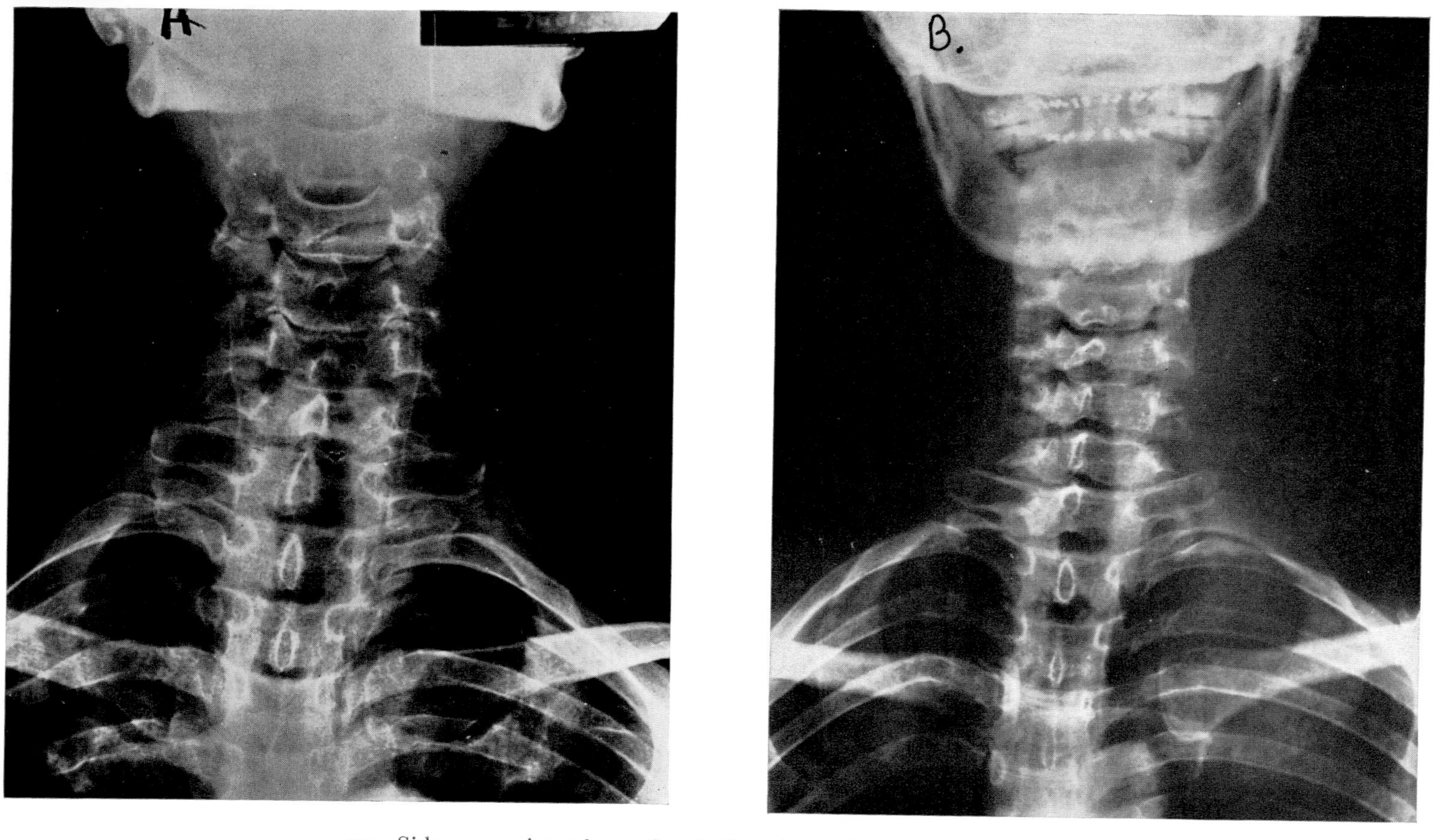

22 Sideways twist at base of neck. See plate 9 (*left*) and plate 19a.

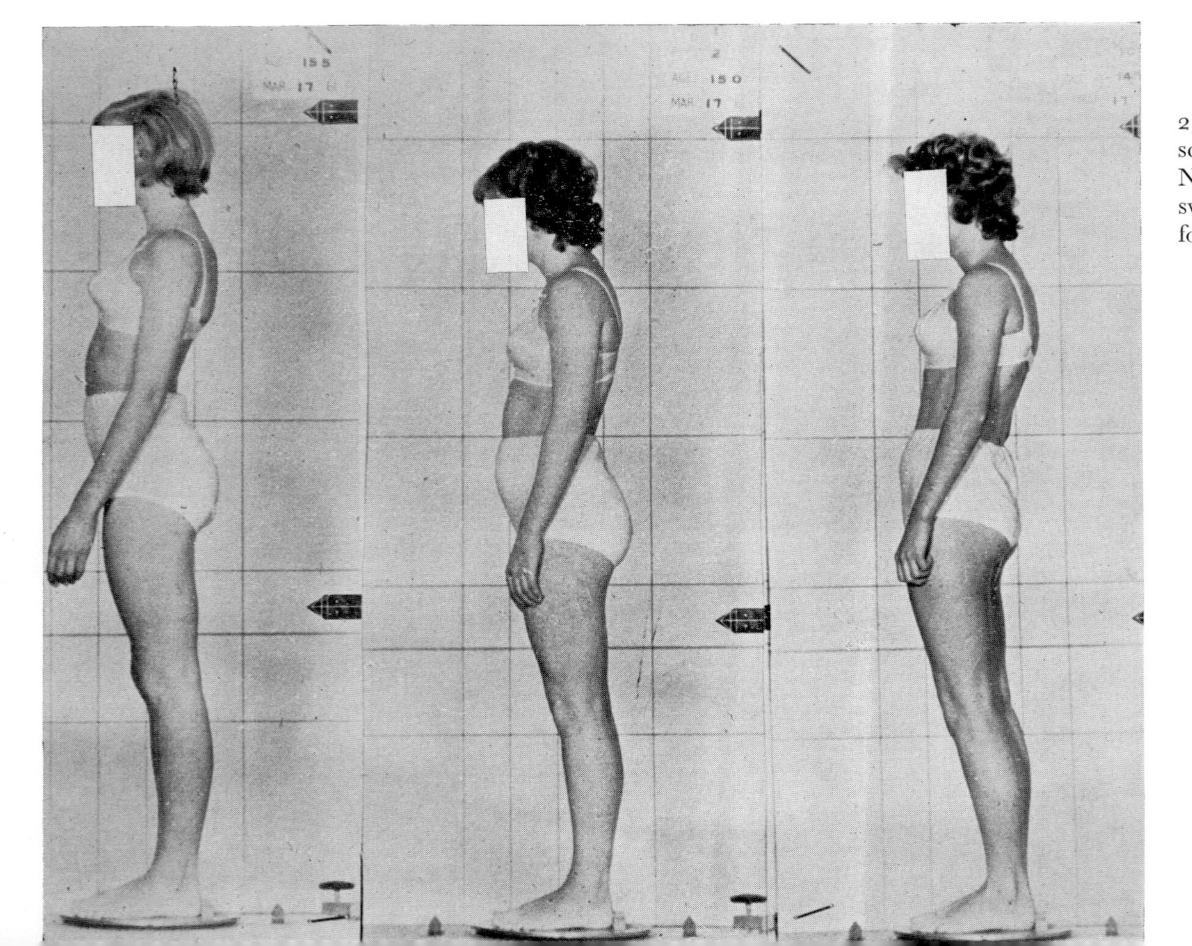

23 Three schoolgirls selected at school as having "good posture". Note multiple severe defects—sway-backs, lordosis, lump formation and head poking.

Whilst we are discussing the head, look at the eyes. Is there a frown? Wrinkles there may be, according to age, but a deeply fixed frown can usually be released a little without loss of social seriousness (if indeed this type of frowning seriousness is ever needed). And the jaws may be held too tightly together—leading in some cases to teeth grinding. The tension will relax if the lower jaw is dropped rather in the drooling manner of the village idiot. When the jaws are closed, the lower teeth should not touch the upper teeth, but should lie just behind them. When the jaw opens it should first drop a very small distance down and then should move slightly forward, in the manner in which a bull dog holds his lower teeth pushed forward. Many jaws are permanently held in an "undershot" position, held back into the throat, not because it was inherited, but because it is held there by habit. And, of course, some people reckon that their chins are big and ugly and hold them in to try to make them look smaller—a dubious cosmetic advantage for which high cost is paid in terms of tension and fixation.

Whilst on the topic of jaws, a word must be said about stammering. In recent years, leading speech therapists have realised and acknowledged how fundamental the Alexander Principle is in their re-educational work, and I have seen many intractable stammerers greatly helped by re-education along these lines. Most of the leading speech training colleges in Britain now know about and use the Principle as a fundamental part of their training.

The Chest

We have already seen that it is easy to develop a sideways twist where the neck joins the back (fig. 20). This is usually accompanied by a throwing of the chest sideways in the opposite direction to the head. The chest (and collar bones in front) which should lie something like figure 21a instead lies as in figure 21b (see page 140).

The collar bone on the side to which it is pushed over may be slightly higher than the other (although *both* collar bones may already be too high because of shoulder tension). The angle between the ribs will be sharper on one side than the other, and in fact the chest may be not so much inflated on that side as on the other, and the cartilage which joins the front of the lower ribs may be felt pushing more forward than on the other side.

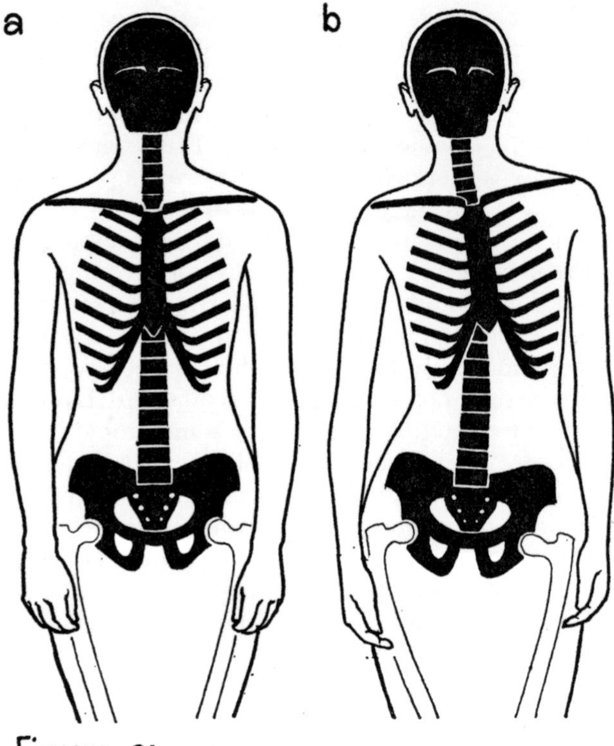

Figure 21

Such chest twists and rotations are frequently unobserved by doctors, and many patients suffer from distressing pain in the chest—sometimes labelled "pleurodynia" or "intercostal neuralgia"—which does not respond to physiotherapy, and leaves lurking doubts that there may be heart or lung pathology in spite of all tests showing them to be clear. (Everything I write presupposes that the usual obvious medical examinations are carried out to make sure that there is no gross pathology. But, of course, even if there is pathology, it can rarely do anything but good to also consider the manner of USE.)

The Abdomen

When looking frontways on, it may perhaps be noticed that one side of the abdomen is straighter than the other—that there is less "waist" on one side than the other. This follows on from the displacement of the thorax sideways (fig. 21b).

The abdominal muscles will be over-stretched and over-straightened on the side to which the thorax has moved, and shortened and "waisted" on the opposite side. And often when this happens, the pelvis on the shortened side will be contracted up towards the chest.

Such twists are often blamed by osteopaths on a short leg, and usually they say this because an X-ray of the pelvis is found to be tilted up more on one side than on the other. There are indeed *some* shortened legs—markedly so in some cases after fractures or polio or arthritis, etc.—but usually an inaccurate measurement of the *true* length of the leg will have been made (i.e., it was measured from the wrong bony points), and the remedy of building up the shoe to lengthen the supposedly short leg will do little to correct an imbalance which stems from a sideways displacement of the neck and chest.

An over-contraction of muscles may also be found in the front of the abdomen: in addition to acting as a form of "muscular armour" which, by continued contraction, attempts to counteract feelings of butterflies or anxiety or sexual stirrings, the over-contraction will occasionally give rise to abdominal pain. The diagnosis of "spastic colon" is very often accompanied by such unnoticed abdominal mis-use, and many people with this distressing condition can be helped, and may avoid needless and fruitless abdominal surgery.

Abdominal laxity and general flabbiness is more often the rule in the middle aged, and to understand this, we need to look once more at our spines and our stance sideways on.

The Spine

Plate 23 shows three secondary school children standing with stomachs collapsed forward. Plate 24 shows a young girl collapsing when she sits. This is the rule, not the exception, in most schools. Alertness at first is made possible by the natural on-going vitality of most of us: but for many school-children, the hours in class are hours of unmitigated boredom, punctuated by bouts of fear and aggression. Added to which the accent on learning to read and to write involves provision of a desk, towards which the eyes and head will tend to drop.* This encourages the formation of an exaggerated hump where the neck joins the chest, plus a slumping of the lower back, and collapse of the rib-cage (fig. 22a).

The slumped head posture becomes habitual so that when the eyes are raised to look ahead this has to be accomplished by an increase in the forward curve in the neck (fig. 22b). But the habit of holding the eyes down may persist until eventually the back of the skull begins to be *held* contracted back into the upper neck in order to look straight ahead (fig. 22c).

In other words the plane of the eyes is altered more by moving the level of the head than by moving the eyes (although they will move a little).

What happens when the child stands up? When it stands up it cannot simply keep a continuous rounded curve down the back, which it had when sitting, since it would produce the impossible (fig. 22d).

So, instead of the good USE in figures 22f, it makes the necessary (but wrong) compensation in the lower back; and this, if we add on the head compensation, gives us figure 22e, the familiar picture of the arched-in lower back, which is so common in school-children over the age of 5, a picture which we saw in plates 9 and 23. And from the back-view we see the sideways scoliotic twists which start in the neck (plate 22) and are compensated wrongly in the chest and lower back.

This combination of lordosis (arched-in back) plus scoliosis (sideways curvature) will persist into adult life and will usually be present when there is chronic back pain (usually it is only in the *acutely* painful back that the lumbar curve is flattened by spasm—a fact which has led some orthopaedic surgeons to try to overcome such flattening and to encourage back-arching). The pathetic picture in plate 25a shows one such patient (who has already undergone a back operation) being encouraged to do just those things which will make his already poor posture much worse. Plate 25b shows a physiotherapist practising this type of exercise.

* A nice verbatim account by Crispin aged nine, "Nearly all the children at school sit with their spines all curled up. That can't be good, can it? Sometimes the teacher tells us to sit up straight and when we ask her why her own back is all curved she says she is relaxing."

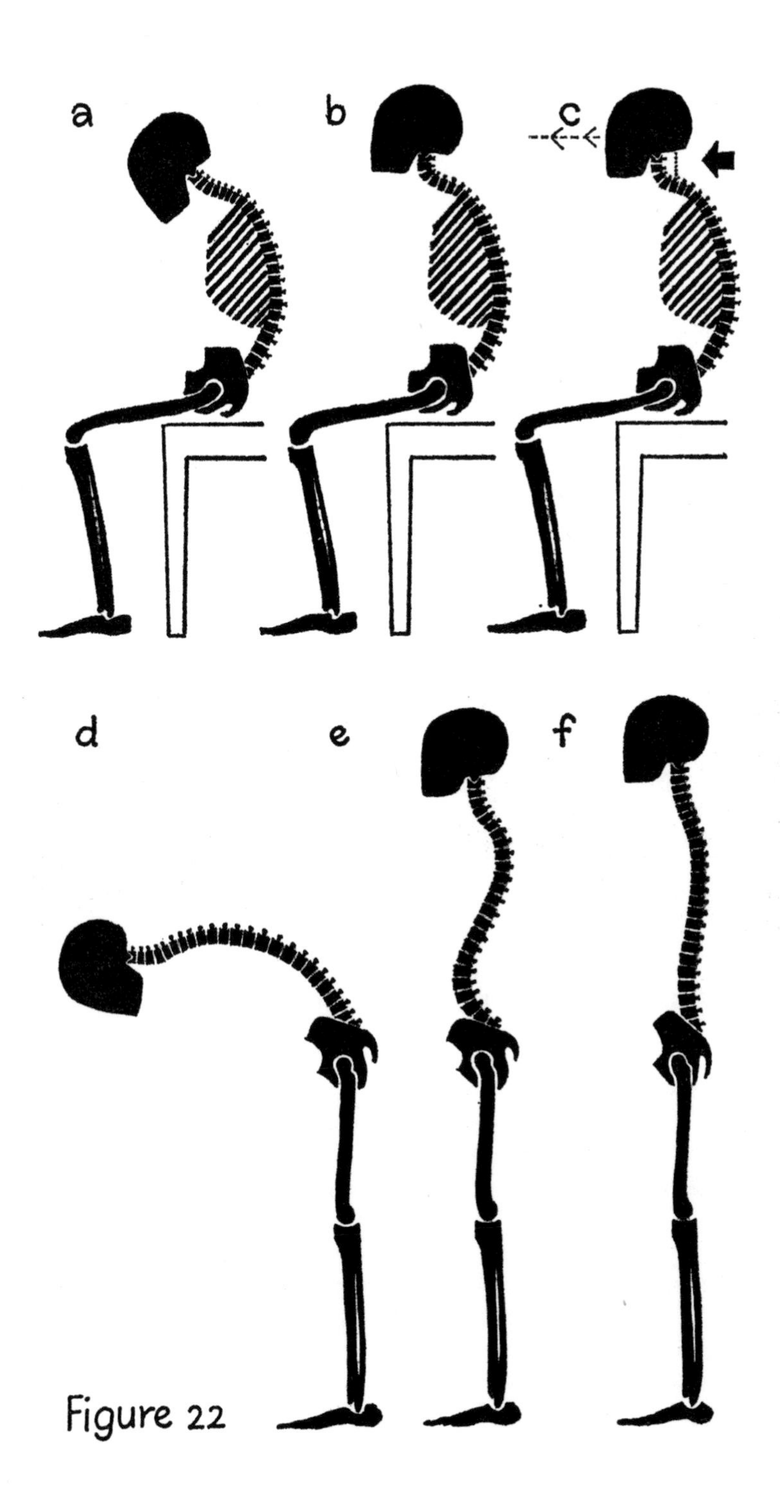

Figure 22

Pelvis and Legs

The pelvis is a most difficult bone to visualise to oneself as figure 23 will indicate.

By far the commonest general mis-use of the pelvis involves pulling the buttocks backwards and upwards towards the lower back—part of the process of arching the back inwards which we have already noticed. But in addition, just like the skull, the pelvis may be tilted up more on one side and it may be rotated back on one side.

The pelvic muscles are extremely complex; on the inside they include the muscles which connect up the lower back with the perineum and legs (fig. 24a and b) and on the outside the small and big muscles which are responsible for standing, walking, running and jumping. One should perhaps make the point

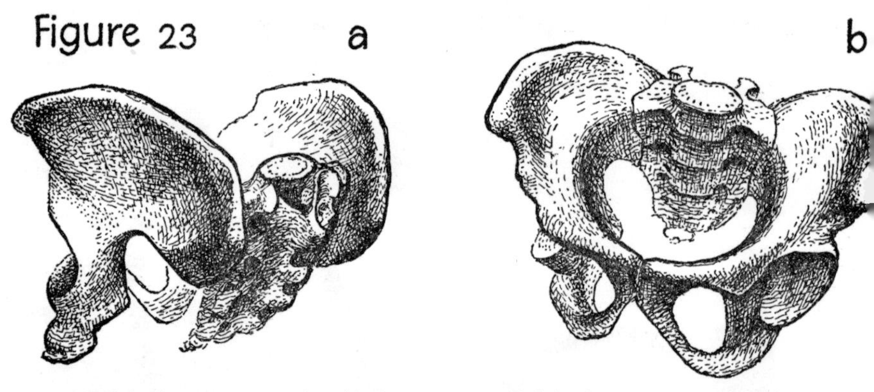

Pelvis three-quarter view back

Pelvis three-quarter view front

again that, when sitting, the knees should never be crossed. If they are crossed it will involve a mis-use of the muscles which connect the lower back to the upper part of the thigh, and the thigh with the pelvis. (figs. 24c and d). A crossing of the ankles is far less likely to involve a mis-use.

Standing and Walking

The following instructions may help the reader to detect faults in the USE when standing and walking.

If you stand with your back to a wall (fig. 25a), with your heels about two inches in front of it and feet about 18 inches apart, you can begin to notice and identify some of your defects.

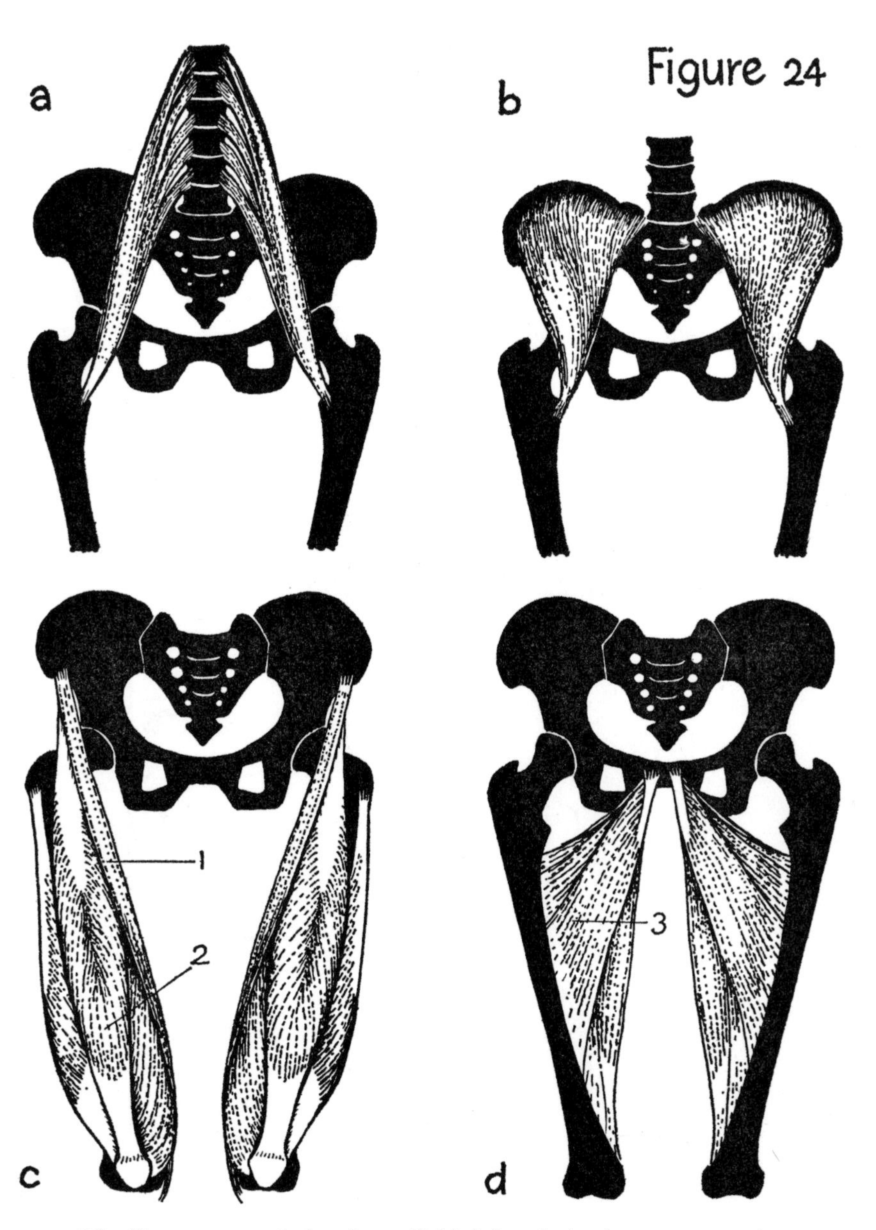

The ilio-psoas muscle has been divided for clarity into two parts (a. Psoas, b. Iliacus). The thigh muscles shown are 1, Sartorius; 2, Quadriceps; 3, Adductors.

Sway your body back to the wall keeping your toes on the ground (fig. 25b). Your shoulder-blades and *your buttocks should hit the wall simultaneously. If you are rotated, one side will hit the wall first; if your pelvis is usually carried too far forward, your shoulders will hit the wall but not your buttocks.*

If the buttocks are not touching the wall, bring them back to the wall. You may notice now that there is a big gap between your lower back and the wall. This gap will disappear if you bend both your knees forward (keeping your heels on the ground) and at the same time drop your buttocks and tip the sexual organs more towards the front, rather than

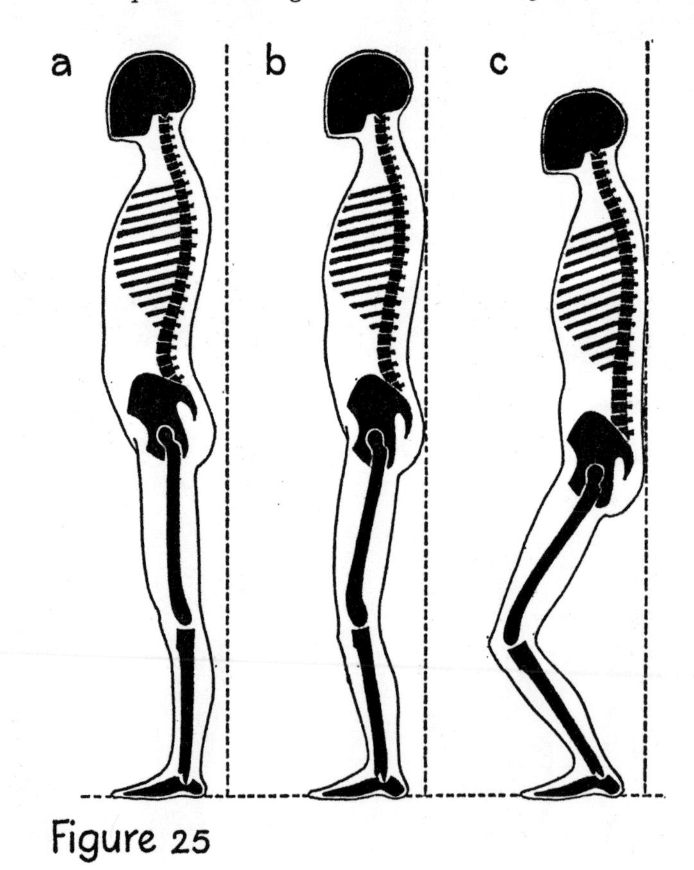

Figure 25

towards the floor (fig. 25c). If you find this position tiring after quite a short time, then you are indeed in a mis-used state! But you will be gratified to notice that your flabby dropped stomach has taken up a slimmer appearance. You may also find that the back of your skull is touching the wall. This is a mis-use—Alexander's basic fault of "head retraction", as described in Chapter 2.

Notice, in this position, whether the arches of your feet are flattened. You can probably unflatten them if, with the knees still bent, you turn your knee-caps outwards, rather as the eye-balls could be squinted out at the sides. This will also tend to correct bow-leg (tibial torsion).

Now slowly straighten your knees but do not brace them back when standing fully erect. The knees should never be braced back when standing, *but should always be slightly bent: the same applies to walking.*

There should now only be a slight arch in the lower back, depending on your particular build and avoirdupois. You can now bring your body away from the wall, keeping your feet where they were. In bringing the body thus forward, the head should lead the movement, not the chest or abdomen.

If you now bring the feet together, you are in a position to detect faults in your walking pattern. Place two high-backed chairs in front of you, figure 26a, and hold them with the tips of the fingers and thumb and with the elbows well out. Begin to walk with the right leg by raising the right knee slightly so that the right heel leaves the ground. But as you do this there will begin to be slight transfer of weight to the left leg to enable the right foot to leave the ground. Many people will find that instead of getting the right foot off the ground by bending the right knee, they instead pull the right side of the pelvis up towards the right side of the chest (fig. 26b). You will detect this happening by the excessive pressure which is felt through one of the hands which is holding the chair. There should be no disturbance of the upper body and arms when this initial movement of bending the knee is made (fig. 26c).

The next stage of walking is simply to go on bending the knee until only the tip of the big toe is left touching the ground (fig. 26d): then as the whole body moves forward, the foot leaves the ground, and should be placed on the ground in front with the heel *touching the ground just before the sole of the foot touches (Fig. 26e). The knee should not be braced back, as in a "goose-step": and if the sole hits the ground first, it will usually have involved too much arching of the lower back. With a heel–toe action it is possible to maintain a use of the lower back in which it is not arched.*

Such a manoeuvre is not intended to teach you how to walk but simply, at this stage, how to detect faults in your walking pattern. A fully integrated pattern of walking would involve very close attention to the upper part of the body, not just to the legs. It is, of course, easier if you

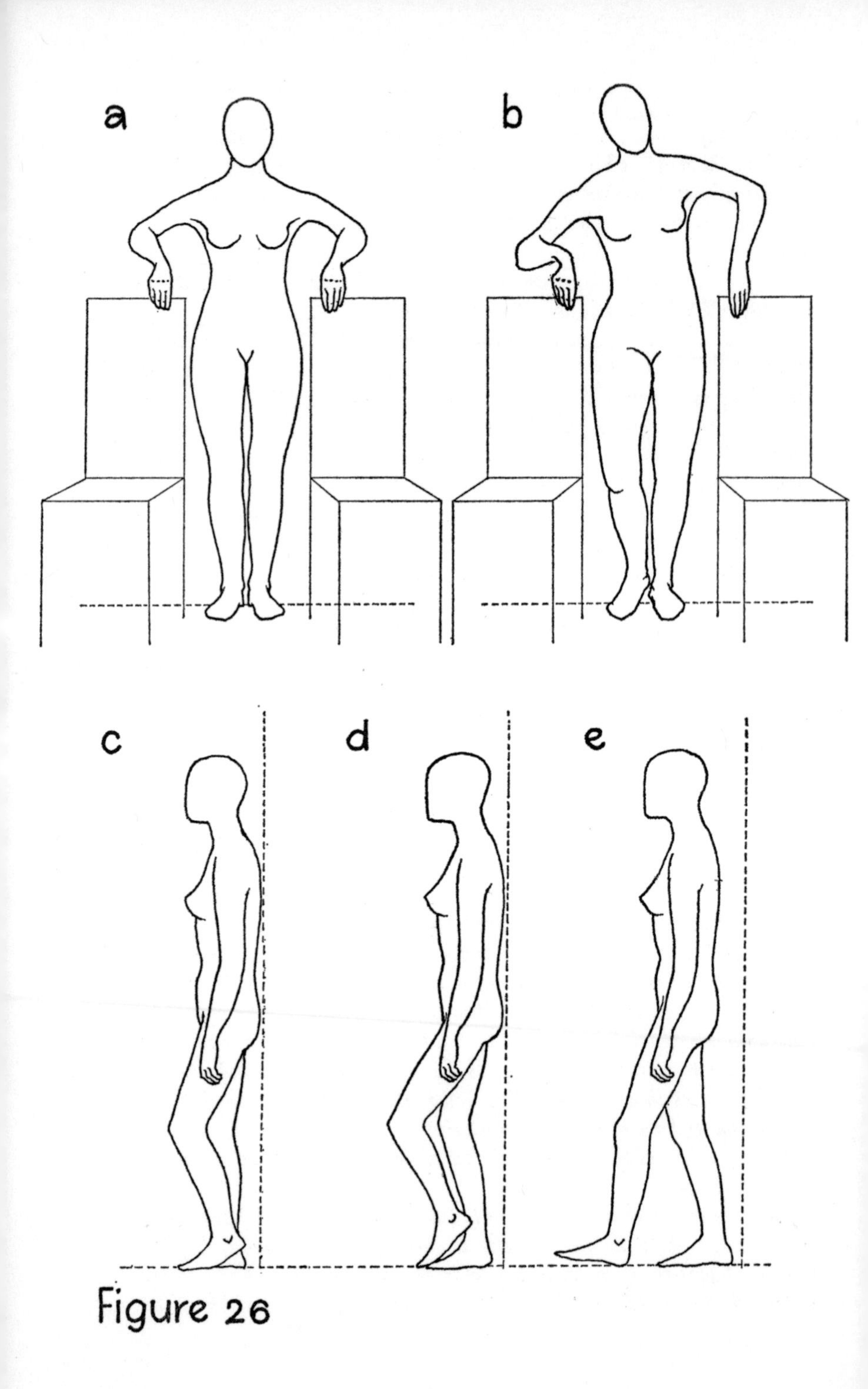

Figure 26

have the help of a teacher who can show you how to hold the top of each chair between your fingers and hands, in such a way as not to create undue tension in the head, neck, shoulders and arms.

Shoulders and Arms

There is no particular reason except convenience for leaving a consideration of shoulders and arms to this point. Indeed, they are, for most people an important site of mis-use, since they are involved in so much of the life of civilised men. Figure 27 shows the muscles which often distort the USE of the shoulder-blades.

In most people the shoulder-blades are drawn up towards the back of the neck during movement and, eventually, even at rest. They will often also be too much pulled together, either because of wrong instruction at school, to "pull your shoulders together": or else because in the sedentary un-energetic life which most people lead, the chest cage is relatively unexpanded and the shoulder-blades—which should lie flat and widened across the back of the properly expanded chest—tend to come together and to become winged when the chest is unexpanded. The bottom end of one shoulder-blade may be felt to be sticking out if the other hand is put behind the back and stretched across to touch the opposite shoulder-blade at the back.

The winged position of the shoulder-blades can be temporarily counteracted by raising the hands and arms forward. This movement (which takes place when holding the top of a chair, as previously described) will usually cause the shoulder blades to lie flat on the back of the chest—a position in which they should lie even when the arms hang to the side. When holding a chair in this manner, the elbows should be turned well away from the body, the cubital fossa (the bend of the elbow) facing the side of the body. This position may feel round-shouldered, but this is because the "hump" is now more obvious and no longer disguised by a spurious squaring of the shoulders.

A pulling of the shoulder blades together is usually accompanied by holding them up too tightly at the back of the neck and hump: and since the upper part of the chest is conical and more narrow than the lower chest, the shoulder blades will wrongly be pulled inwards towards the hump as they rise up over the conical chest. Clearly such tension has to be released by a slight dropping and widening of the shoulder blades: but most people find it difficult to do this without slumping their lower back at

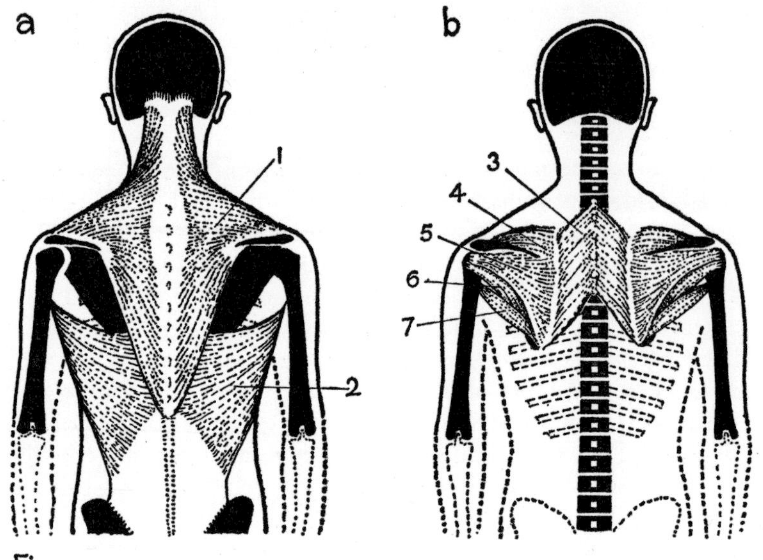

Figure 27

1. Trapezius; 2; Latissimus Dorsi; 3. Rhomboids; 4. Supraspinatus; 5. Infraspinatus; 6. Teres Minor; 7. Teres Major.

the same time. So not only do the shoulders have to be released and widened, but the back has, at the same time to lengthen (without arching) and to widen to support the shoulders; a process which is facilitated if the chest cage is expanded to widen across the back. Plate 26 shows such a widening, with chest expansion. Plate 27 shows the slow improvement in the right (but not left) shoulder tension.

A great deal of emotional tension is expressed in the shoulders. Dorothy Tutin told me that she played Joan of Arc with the shoulders slightly raised and fixed to give a feeling of defiance. A whole range of most subtle emotions are manifested in the shoulders—shrugs, aggressive threatenings, or resignation and nostalgia from, say, a slight release of the upper chest muscles as they insert into the upper arm in the armpit.

The Arms

The bend of the elbow should not be facing forwards when standing: indeed, the upper arm should turn slightly inwards,

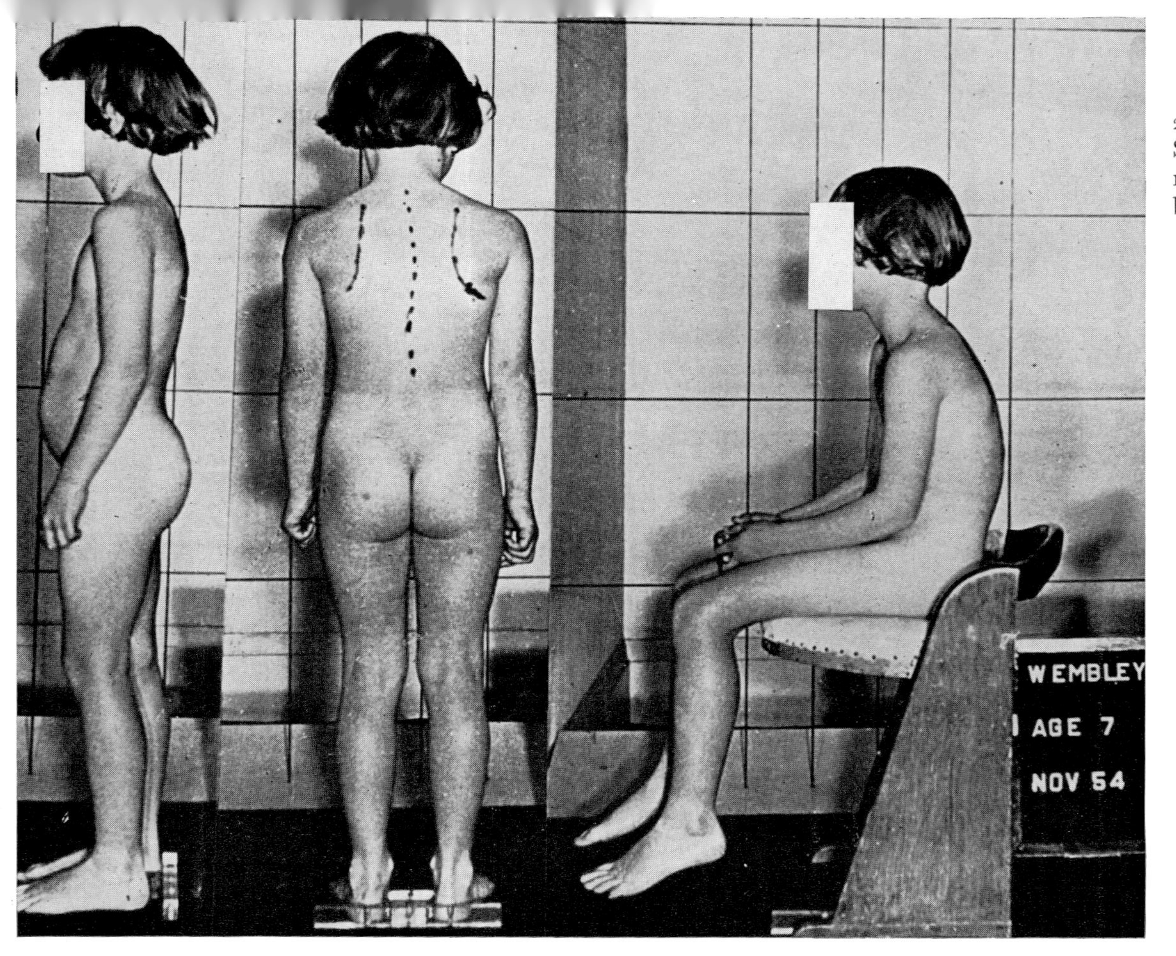

24 Typical schoolchild posture. Spine curved as in figure 7, neck dropped forward, head back.

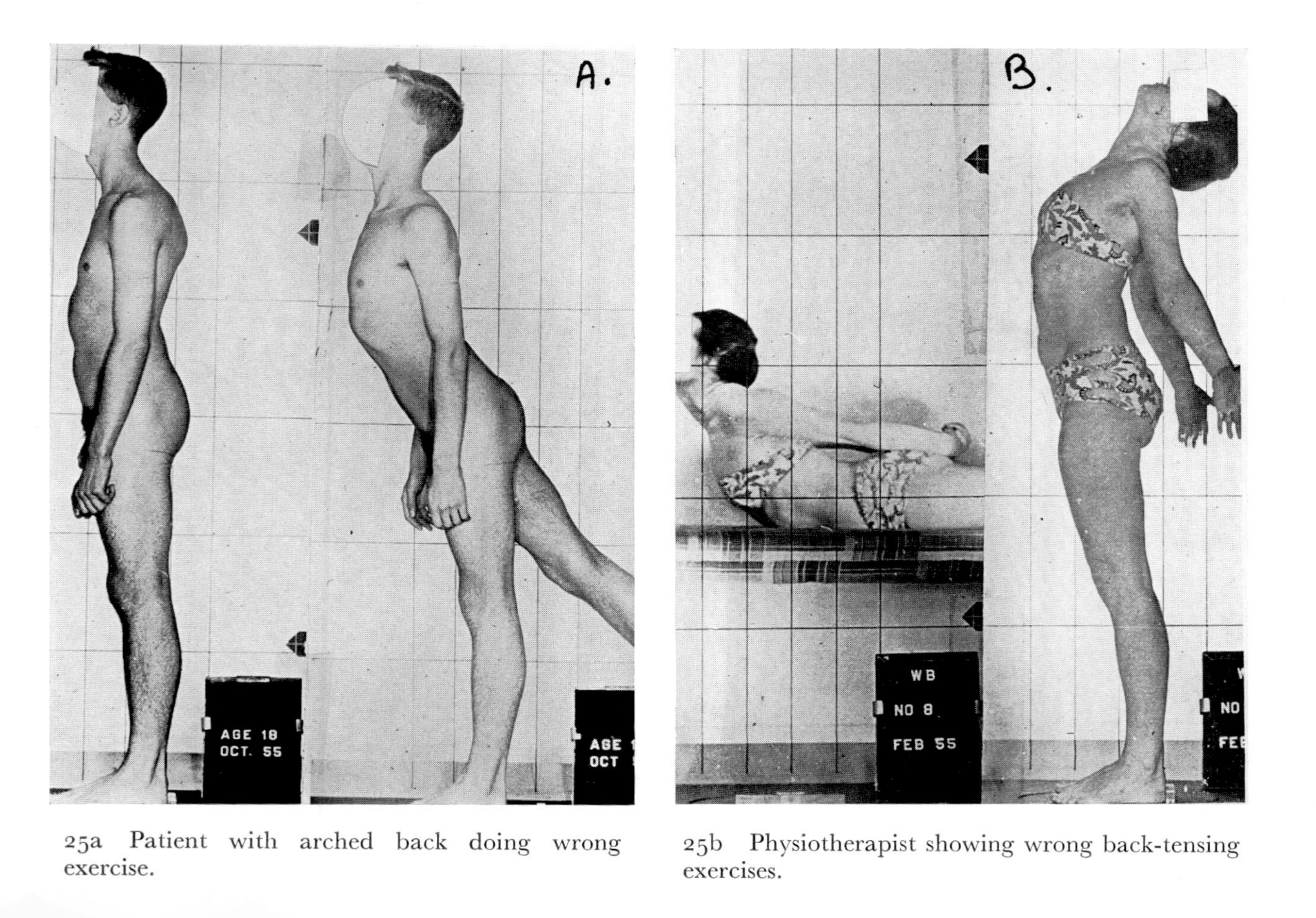

25a Patient with arched back doing wrong exercise.

25b Physiotherapist showing wrong back-tensing exercises.

26 (*right*) Widening of shoulder-blades apart. Note loss of skin fold in b after re-education.

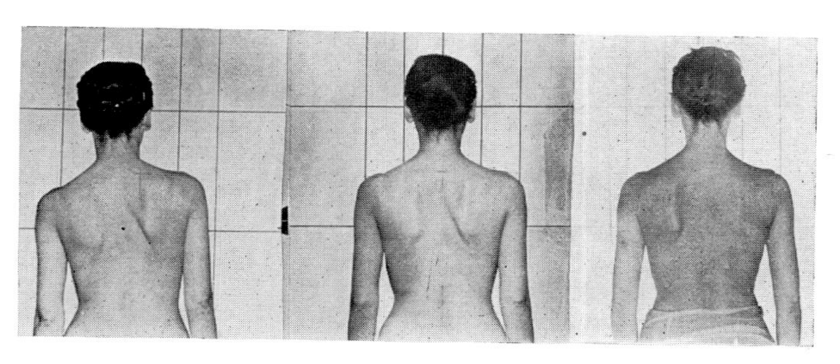

27 (*above*) Back straighter and right shoulder widening after re-education.

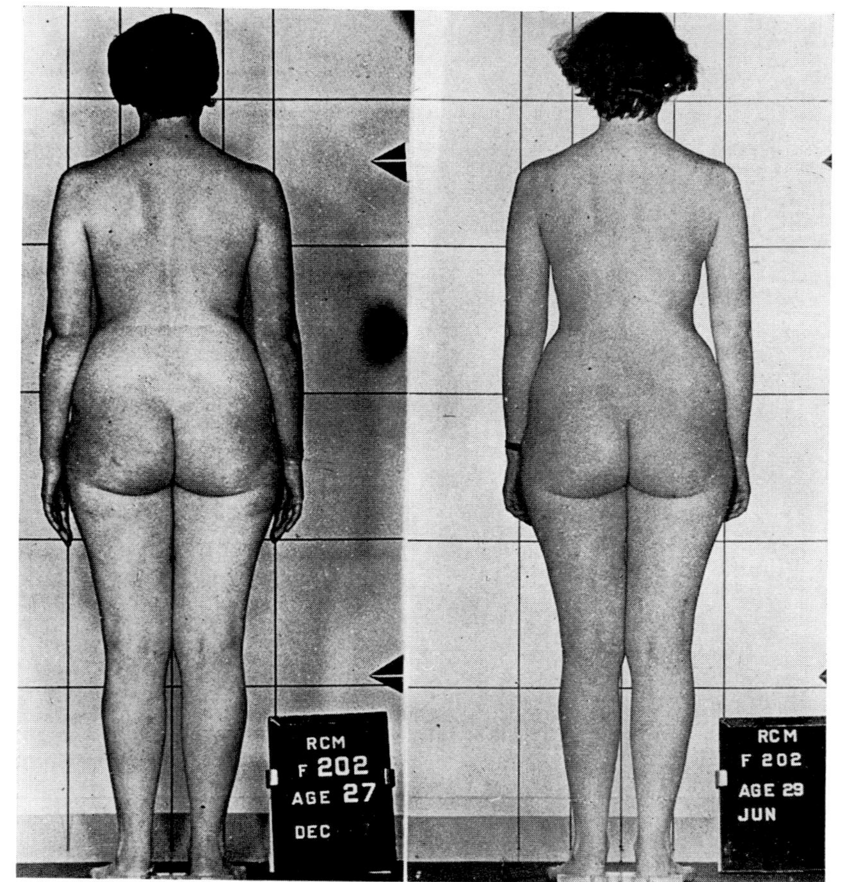

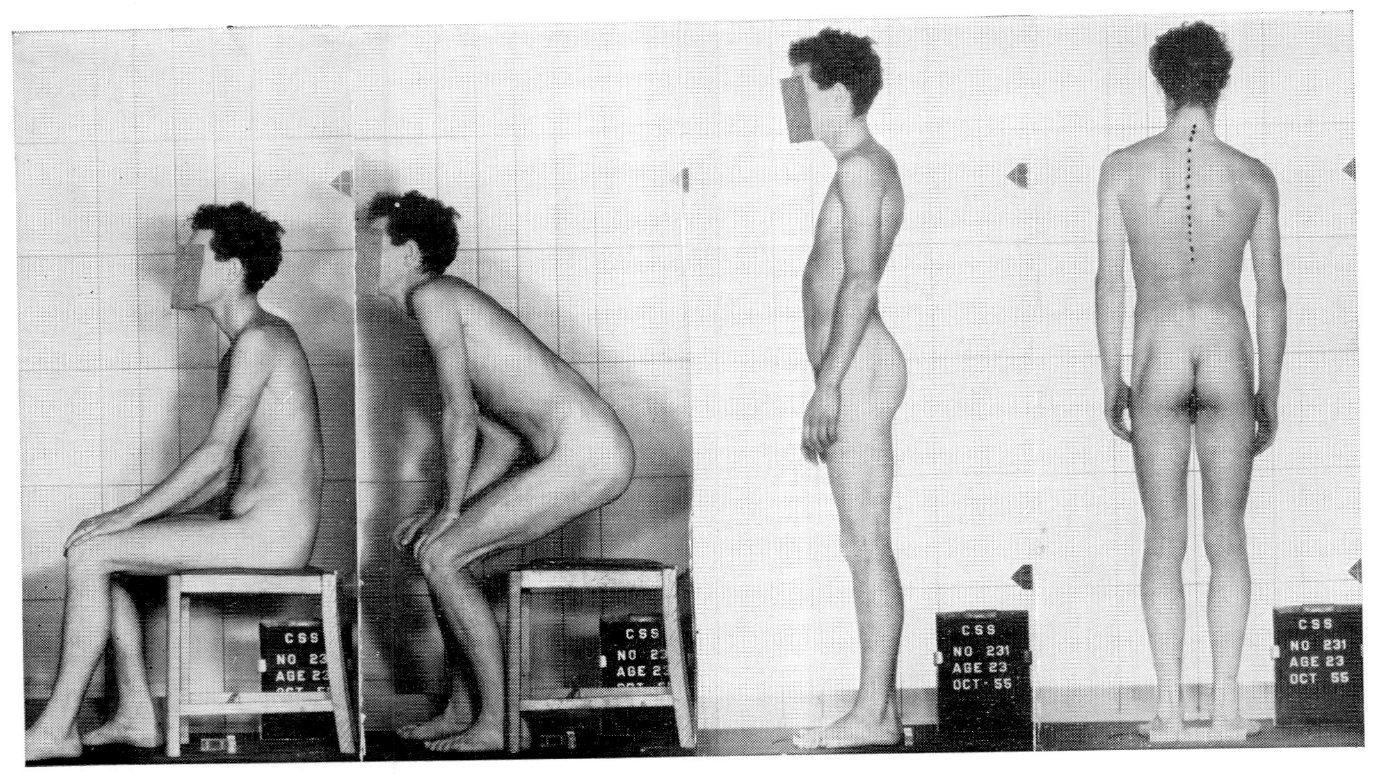

28 Most of the possible defects! Slump, head retraction, lordosis, scoliosis, shoulder raised, right hand dropped, tibial torsion.

so that the elbow turns slightly out and away from the body. If the shoulders are collapsed, the upper arms will also usually be held too close to the body.

The front of the elbow is a place where most people hold themselves far too fixed, so that the fore-arm may be flexed, at rest, too much towards the upper arm. We use our arms for many of our activities of daily living, and we gradually come to adopt a resting position in which the fore-arm is too clenched, the elbow a little too bent, and the hand not sufficiently straightened out. The line from the inside of the elbow (which can be drawn down, via the inner border of the wrist, to the end of the thumb) should be almost straight, so that the fingers, at rest, can be straightened and turned slightly away from the thumb. Driving a car, playing a musical instrument, handling tools and objects of all kinds, can be most economically carried out from this resting position, which will also involve a broadening of the palm of the hand and a separating of the fingers. One should endeavour to adjust the height and position of desks, musical instruments, display-panels on work-benches etc., so that such good use of the arm and hand is not hampered. And in writing and typewriting, there should not be too much deviation from the use of the shoulders, arms and hands which I have described. And of course in such all too frequent medical conditions as the stiff "frozen" shoulder, tennis elbow, and "tenosynovitis" of the wrist, it is of paramount importance to establish correct habits of using the whole upper limb, and to ensure adequate postural support for it from the trunk. Likewise in various games and athletic activities, it is essential to learn a basic resting use of the arms and shoulders: a consideration of such mis-uses may give a clue to persistent putting-errors or slicing of the golf ball, or patches of erratic serving and smashing in tennis, and so on.

This brief account of some of the more obvious mis-uses scarcely touches the fringe of the problem. It is easier to demonstrate these things on living persons—it took a long weekend at an adult education college in which I gave six lectures, demonstrated for four hours on local school-children and individually analysed all those taking the course, before I reckoned I had established all the structural points I wished to make. Plate 28 shows most of the common defects—the slumped sitting position, head pulled back when standing-up, dropping of the shoulder and arm with thorax displacement, neck dropped forward, prominent hump, arched back, knees too braced back. And all in the same young man, aged 23.

The Incidence of Mis-Use

A given individual not only at the macroscopic level but much more importantly at the microscopic level will create his own permutations and combinations out of the many alternative mis-uses. A neat classification of mis-use is not at present possible, but in spite of this, it is possible, using broad categories, to get some idea of the extent of the problem.

Various physical education colleges and local education authorities have, over the years, allowed me to carry out studies on their students and school children, and I was the medical member of the National Committee on Movement Training a few years ago, in which we attempted to assess the physiological effects of various types of physical education.

The pro forma, figure 28, has proved a useful guide in assessing a given person's defects, and it has been adopted by educational authorities in their remedial work. It can be scored on a simple basis of one, two, or three marks according to the severity of the defect. Most of these "posture studies" have been carried out with the help of Professor Tanner from the Institute of Child Health. His book *Growth and Adolescence* (*op. cit.*), will give an indication of the detail in which such studies are carried out.

Surveys were carried out at some of our leading physical education colleges, where a high physical standard is required at entry, and which draw on some of the best athletes and games players in the country. It can be reasonably claimed that this group of students have applied themselves from an early age to the development of their bodies, and they are destined to become physical education teachers all over the country.

When analysing their faults by photography, a definite pattern appeared (fig. 29) of well defined categories—those scoring 0–3 who have excellent USE, those scoring 4–5 who have some slight defects, those between 6–9 who show severe defects, those between 10–14 who show very severe defects and those over 15 who show really gross deformity. In a group of 112 physical education students (fig. 29b) the majority, 62%, showed severe defects, 11·5% showed slight defects and 26·5%

Region	*Faults*	*Score*	*Faults*	*Score*
HEAD	Poked		Tilted	
	Retracted		Pulled Down	
SHOUL-DERS	Raised		Rotated	
	Dropped		Pulled Together	
PELVIS	Tilt		Forward Carriage	
	Rotated		Gluteal Asymmetry	
SPINE	Scoliosis		Thorax Displacement	
	Kyphosis		Lordosis	
STANCE	Hyperextended Knees		Forward Inclination	
	Internal Rotation Knees		Symmetry	
TENSION	Specific			
	General			

Figure 28 Form for recording postural faults

showed very severe defects. Similar figures were obtained from groups of male, (fig. 29a) and female (fig. 29c) drama students.

On the basis of such studies it is clear that whatever methods are being used in our schools, the end-results even in the best students are not good. The idea that a healthy natural outdoor

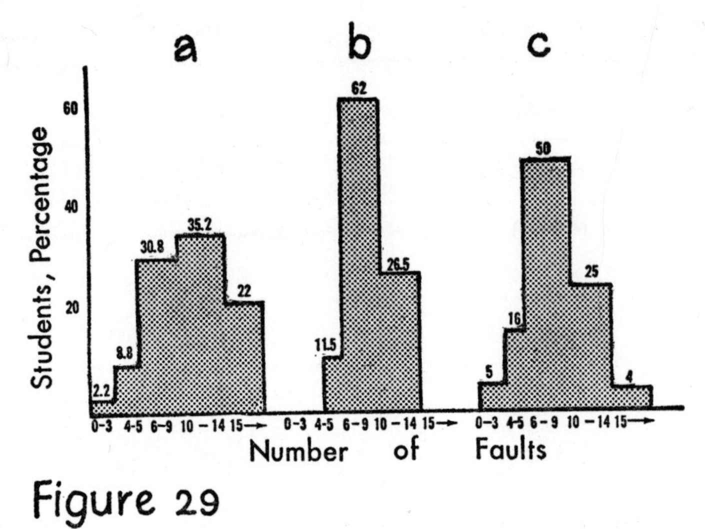

Figure 29

life with plenty of fresh air and exercise will ensure a reasonably good USE is simply not true—we see how high is the incidence of defects in physical education students. However, even if it were true, the main problem would still be how to establish a USE which would stand up to the strain of living in a civilisation in which the healthy life may not be easily available.

The figures of wrong USE in students are alarming but equally alarming were the figures which were obtained by the National Committee for Movement Training of the USE of children in secondary schools. This study was carried out in schools in Hertfordshire and, in addition to a whole battery of physiological and psychological tests, I carried out a survey of postural and tension defects.[40] It was encouraging that the committee's comments at the end of the study were: "The main hope for the future seems to be the kind of measurements and ratings that Dr Barlow and his colleagues can make."

In a world survey of physical education methods, which was carried out in the UK, in the US, Australia, and the USSR, the report was concluded by an account of the work which Professor Tanner and I have carried out, and the report quoted me as follows: "At present, Physical Education training does not leave its pupils with either the knowledge or the desire to maintain healthful activity in advancing years. The problems of adult deterioration under civilised conditions is far more important than the problem of providing healthful outlets for the young. Physical Education has failed unless the adult both desires and is able to maintain good USE throughout his life."

The report concludes: "Both these doctors are in touch with members of the Physical Education Association, and it is hoped that their interest and their wider opportunities for research into the problems of physical development and movement may guide physical educators in their educational work and be of special help to those responsible for corrective physical education."[41]

It is difficult, unaided, to learn the Principle, and accordingly it is vital that this knowledge should eventually be available at the school level. Just how it is to be fitted in will be a matter for individual headmasters. It is tempting to feel that the Alexander Principle will be too strong a medicine for some schools at present. However, through the patience and skill of certain Alexander teachers, who have quietly worked to prove to other school staff just how valuable their contribution can be, it does seem now that the Principle can be acceptably incorporated into many schools and that this process will be mutually helpful on both sides.

In recent years Alexander teachers have been taken on to the staff of four colleges—the Royal College of Music (where most of my early research was done), the Royal Academy of Dramatic Art, New College of Speech and Drama, and Guildhall. The Inner London Education Authority has recognised the Alexander Institute for provision of major county awards—a move which may do something to satisfy the huge increase in demand for trained teachers. But, in the main, most Alexander training is still carried out by individual teachers who work privately. It has often been the case in the past that when a new educational need has arisen, it has at first had to be dealt with in the private sector, rather than by the state. It is only when the evidence has become incontrovertible that the institutions join in.

The evidence now is quite incontrovertible. We are witnessing a widespread deterioration in USE which begins at an early age, and which present educational methods are doing little to prevent. Most people have lost good USE by the time they are past early childhood. Not all the time, but in most of their activities and when they are resting. Nobody bothers about it because nobody notices until the defects have become severe. Many family situations are bound to produce tensions in children, but it is difficult enough even for the lucky ones whose parents provide a balanced environment. It is difficult for children to "keep their heads" when "all around" them they are surrounded by people who are monstrous monuments of mis-use.

Alexander devised a method for teaching improved habits of USE, and this has become widely known as "The Alexander Technique". The next chapter will explain some of the concepts that underlie his technique and will describe procedures carried out in an Alexander lesson.

Chapter 10

TEACHING THE PRINCIPLE

Some come in hope and others come in fear,
Diverse in shape the multitude appear.

POPE

ALEXANDER INSTRUCTION IS, at first, an individual matter, one-to-one: it cannot be skimped. Unless it is detailed, it is nothing. The Alexander teacher will have his subjects with him for between half an hour and an hour, during which time there is a continuously absorbing preoccupation with the development of new USE. For most people this will involve at least fifteen sessions: for many it will involve very many more. Relatively stereotyped procedures of physiotherapy or manipulation may produce a greater turn-over of patients, and with less effort. This can never be the case with Alexander instruction. A teacher can only deal with a comparatively small number of people in a working day.

Learning a better general USE is no different in kind from learning any other specific skill, but with or without a teacher, it will involve working at two simultaneous levels—a mental "labelling" level, and a psycho-physical "experiencing" level. To illustrate these two levels, let us think of someone who is blindfolded and who is trying to make out what an object (A) in front of him is:

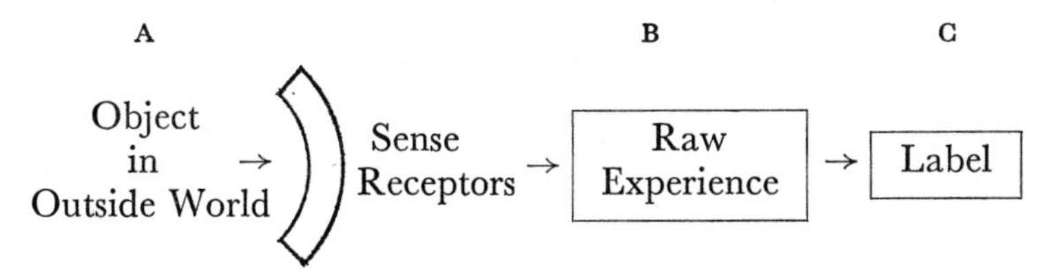

His sense-receptors may tell him first that it is heavy, smooth, cold, rounded, etc.,: he is getting from the senses a "Raw" unformulated experience (B). Very soon he will think he recognises it and may give it a label (C): a "jug", perhaps.

The labelling process, C, involves discrimination and recognition according to what we know already about jugs, etc. The label is *not* the experience: the word is not the object. The two levels must not be confused, although one leads to the other.

An animal in the forest has, presumably, a fairly simple labelling process: if it is a monkey, which is in ever-present danger of being eaten by a tiger, the important thing is to be able to detect the tiger before it actually arrives. A scent on the breeze—level B, is associated in a flash with the thought, level C, of a tiger, and this stimulates an immediate response—probably shinning up a tree with all the other monkeys. However, it would be useless for a well-meaning stranger to shout "Look-out! Tiger, Tiger!", because the verbal label "Tiger" would mean nothing, whereas the scent on the breeze would mean everything. Words and ideas by themselves, are not a sufficient form of education in anything which involves the senses, and they can only become effective when they have been linked to a raw experience by a learning procedure. Thus, in time, a monkey might be trained to get up a tree whenever someone shouted "Tiger".

If a new USE is to be learned, it is necessary for many new experiences of USE, level B, to come to be associated with new labels, level C: and the new labels will, in time, come to induce the improved USE. In human beings, the new USE-label cannot be as simple as that of the conditioned animal: but a simple association must first be set up between the names of various USES and the actual USES themselves. A grammar of the body has to be learned—a grammar not to be triggered into action by some external cry of "Tiger, Tiger", but by being projected personally as and when one wishes.

Grammar

An electrical recording which has been obtained from a muscle during re-education shows this procedure clearly. Figure 30 shows a recording from neck muscles which are being trained to release, whilst, at the same time, the patient repeats to himself a verbal direction "neck release". In (a) there is much initial tension in the neck: in (b) the tension becomes momentarily less, but returns as soon as the teacher stops his gentle adjustment of the head. In (c) the improved balance is obtained sooner, but still it returns when the teacher stops his adjustment and likewise in (d). In (e) the patient is now able to maintain

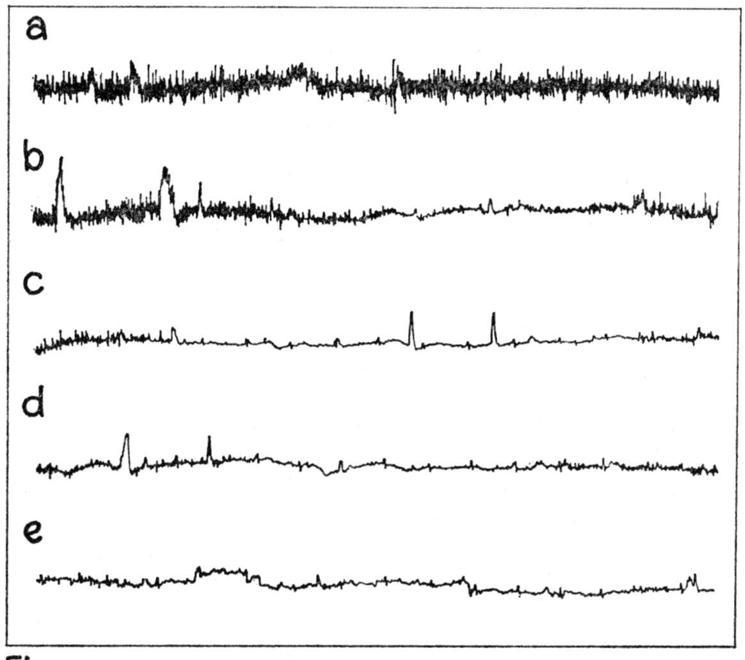

Figure 30

the state of lessened tension and, in time, will be able to evoke this state simply by running over the "orders".

A sequence of such verbal directions is taught whilst a better tensional balance is obtained all over the body; the sequence is designed to scan the body in serial order, much as a television camera scans its object or as one scans a telephone directory in search of a number. The sequence of directions thus provides a model with both spatial and temporal co-ordinates. Such a sequence fulfils the function of checking the development of too much tension and of restoring a resting state when it has been disturbed. If kept in mind during performances, it will ensure that deviations from the resting state are not excessive.

However, to learn such a "Body-Grammar" is not in itself enough. Two other factors are needed. First, the subject must learn not to "End-gain": and this will require what Alexander called "Inhibiting".

End-gaining

Most people prepare for action by creating unnecessary muscle tension and one of the most interesting and original ideas which Alexander put forward to explain such unnecessary tension was his concept of "End-gaining". In order to understand what takes place in an Alexander lesson it is essential that this concept should be understood.

Alexander meant it to apply not only to wrong methods of teaching and learning a new USE but on a much more general basis. Briefly, end-gaining means the habit of working for ends, targets, goals, results, without considering the means: without ensuring that the means we employ won't produce too many harmful by-products.

End-gaining shows itself in the form of over-quick and over-energetic reactions. Targets—when we live by the end-gaining principle—have to be reached as soon as possible, so that yet another target can be achieved. Not only big goals and achievements: but small actions like turning the tap on, picking things up, swallowing food, or interrupting people.

In end-gaining terms, a successful life is one which achieves more and more goals, and the devil take the hindmost. These goals may either be personally selected or they may be the ones which we have been encouraged by our society to select: but whether they be the goals of the "organisation man" or the lonely pioneer, "end-gaining" implies that proper consideration is not being given to the USE involved in gaining the end.

This concept of end-gaining was taken up with avidity by John Dewey, the American educational philosopher,[42] who saw it as a way in which children could become more interested in what they were actually doing than in the pat-on-the-back which they hoped to get from exam success or from being top of the form. It was also taken up by Aldous Huxley who wrote a book—*Ends and Means*[43]—about it:

> We are all, in Alexander's phrase, end-gainers. We have goals towards which we hasten without ever considering the means whereby we can best achieve our purpose. The ideal man is one who is non-attached. All education must ultimately aim at producing non-attachment.

Modern interest in meditation has led some writers to see in Alexander's "end-gaining" the same fault which they themselves are trying to eliminate. For this reason some of the writing about Alexander has had a mystical flavour, not to say a sentimentality. Experiences which can come from what Alexander called "inhibiting" such end-gaining have been seen by some people as more important than any of the other effects of Good USE.

Inhibiting

"Inhibiting" as Alexander suggested it, is not to be confused with repression and unresponsiveness as understood by the psychotherapists. Alexander had a fairly simple stimulus/response psychology and his behaviour diagram went something like this:

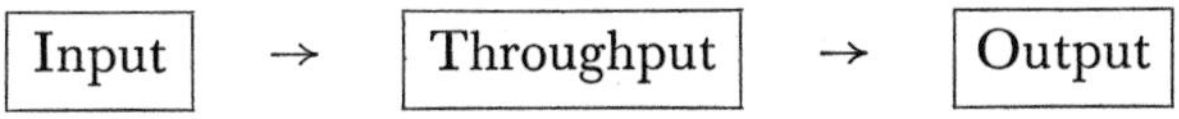

Input	→	Throughput	→	Output

He rightly saw that end-gaining was a reflex action which tended to by-pass the reasoning brain and that most end-gainers are reacting automatically on an input/output basis, so that activity is directed towards satisfying the input as soon as possible, whether or not the habitual way of doing this is appropriate. He accordingly insisted that on the receipt of a stimulus, there must be an "inhibition" of the immediate muscular response, so that by "throughput", there could be adequate preparation for the succeeding activity. Such inhibition became a cornerstone of his re-educational methods.

This is an important and basic observation; but the snag which may arise from a rigorous determination to refuse to react and to register sensations is that curious states of disorientation may be induced in this way and, like the philosopher who stood on a river bank watching someone drown whilst he tried to decide whether human life was valuable, the use of inhibition has led some Alexander adherents in the past to a state of passivity in which they prefer not to respond at all in case their dystonic patterns should reappear. Such "sensory deprivation"—which may come from over-zealous inhibition—must clearly be used with care. This snag, however, disappears if the period of inhibition is seen merely as a stage of preparatory choice in which the eventual muscular USE can be decided on—a stage which leads on to activity or to a state of freedom whilst at rest.

There will be much to say about "inhibiting" in Chapter 11 (Learning the Principle). Consideration can now be given to the Alexander Technique and a specimen Alexander "lesson".

The Alexander Technique

The Alexander Technique, briefly, is a method of showing people how they are mis-using their bodies and how they can prevent such mis-uses, whether it be at rest or during activity. This information about USE is conveyed by manual adjustment on the part of the teacher, and it involves the learning of a new Body-Grammar—a new mental pattern in the form of a sequence of words which is taught to the pupil, and which he learns to associate with the new muscular use which he is being taught by the manual adjustment. He learns to project this new pattern to himself not only whilst he is being actually taught but when he is on his own.

This procedure is *not* a method of manipulation in which the subject is a passive recipient and goes out none the wiser about how to stop himself getting into a tension-state again. It is a method by which he is taught to work on himself to prevent his recurrent habits of mis-use, and by which he can learn to build up a new use-structure.

It is *not* a form of hypnosis, by which the mind is conditioned to obey commands which some other person has planted there. It could perhaps be described, in its initial phase, as a *de*-conditioning, since it aims to teach the pupil to recognise when he is making faulty tension: but it is not akin to the deconditioning procedures of behaviour-therapy, in which, say, the tension of a writer's cramp is punished by giving an electric shock every time the pen is held wrongly. Instead, it involves throughout a conscious attention and learning by the patient, and no adjustments are considered useful unless they can be built up into a new conscious body-construct, to be used consciously by the patient afterwards.

It is *not* a form of relaxation therapy. Certainly the pupil will learn to release tension which was previously unconscious, but in all probability the pupil will be expected to replace unnecessary tensions by additional work in other muscle groups which previously had been under-developed. Many people, for example, having picked up the bad habit of slumping and crossing their knees, will find that they will have to put more work into their lower back and thighs if they are to release excessive tension which they have been making in their neck and shoulders. The over-tension of their neck and shoulders will be replaced by more tone in the lower back and thighs; but, of course, each of us has his individual patterns, and each of us will have different patterns of wrong muscle tension which need to be redistributed.

Exercise Therapy

How is this learning to be brought about?

Not by "exercises" as they are ordinarily understood—the "stretching" and "strengthening" exercises of physiotherapy and physical education are valueless in dealing with this problem.* It might be argued that from its original use (*ex-arcere* meaning to "break-out" from a "shut-in" state) the word exercise would be appropriate for a procedure which seeks to release dystonic patterns. But since the hardest problem of all in such re-education is to stop people "doing" too much, the word "exercise" is probably best confined to such phrases as "Directive Exercise". The word "Practise" is not so suspect, so long as it is not divorced from a real-life context.

Defective Awareness

Even so, it might be thought that once a proper USE-diagnosis has been made, learning should be easy. Given that people are interfering with their USE in a certain way, all that is necessary, surely, is to show them what is right and get them to practise it: then all will be well.

Unfortunately the problem is not so easy. When people have grown accustomed to a certain manner-of-use, no matter how twisted or crooked it may be, it will have come to feel "right" to them even though it may also be producing pain and inefficiency and clumsiness. Our sense of "rightness" is a very precious possession to most of us—it is, after all, the outcome of our whole experience of living to date, and our whole nature is bound up with the pattern of movement and posture which we have developed. Indeed, our whole memory pattern is closely tied-in

* W. Barlow, Proceedings of the Royal Society of Medicine, *op. cit.* Over a twelve-month period, students who were given such "exercises" had developed more postural defects than they started with.

with the substratum of muscle-tone which underlies our USE. A sense of the space co-ordinates of our postural system pervades all our behaviour. We *are* our posture.

This means that, the moment we try to carry out a basic re-education of USE, we very rapidly run up against our attachment to our old feeling of ourselves. We can be shown in detail what are our defects, and most of us will readily admit that they need altering, and that our general feeling of ourselves must be inaccurate for us to have got into such a mis-used state. In spite of this, however, most people try to correct their mis-uses by deliberately taking up some new position which they think is what is now required. In the process they will only create yet another set of dystonic patterns: and it will not be long before they revert to the old habitual pattern, particularly when they start to move. It may be recalled that out of my group of 105 young men (Chapter 2) only 11 out of the 105 were able to alter their habits at will and then only at the immediate moment of supervision.

The Alexander Lesson

Alexander himself has described how he managed to teach himself a new body-construct, and to associate this with a new manner-of-use.[44] But this is a laborious business on one's own, and, in common with most skills, it is all far easier with the help of a good teacher.

It may be helpful to describe an Alexander lesson, although of course each teacher will have worked out his own ways of presenting the necessary information as clearly as he can.

An initial examination by the teacher will have indicated that there are certain forms of wrong USE which are deeply established. These uses will be showing themselves in the disposition and alignment of the bones of the vertebral column and limbs, and they will be showing themselves in a disposition to react with the muscles in certain habitual ways. It is taken that the pupil has been sufficiently persuaded by this diagnosis to accept the type of instruction which is being given.

The pupil will be asked to lie down on a fairly hard surface—the usual medical or physiotherapy couch is a good height. The head will need about 1 inch of firm support under it, although up to 3 inches may be needed if the Hump has become fixed and very bent forward. The pupil will be told not to do anything—in other words to "inhibit" any reflex movement which tends to take place when he is handled or moved.

The teacher will place both of his hands at the sides of the neck and will ask the pupil to say to himself the words, "NECK FREE, HEAD FORWARD AND OUT". He will perhaps explain that "FORWARD" means the opposite of pushing the skull back into the support under the head, and that "OUT" is the opposite of retracting the skull, like a tortoise, into the hump and the chest. (When standing or sitting, the instruction may be "FORWARD AND UP", which amounts to the same thing.) The teacher will repeat these words as he gently adjusts the head in such a way as to release neck tensions which are preventing it from going forward-and-out. These neck tensions will be manifold and different in each one of us. The neck X-rays (plate 18) showed just a few of the sorts of contortions into which we can get our necks and heads. All such contortions are gently corrected—to some extent—by the teacher as he takes the head forward-and-out. Obviously the tensions and distortions of a lifetime won't come undone in one magic moment: and equally obviously, the somewhat improved balance which is initially attained in this way cannot be taken to be the one final and right answer. Over a period of days and weeks, the pupil will gradually become familiarised with an improving and changing use of the head and neck.

This familiarisation with a new USE would not, in itself, be of much value. It is also necessary for the pupil to learn how to maintain such an improving use when he starts to react.

At first, the teacher will present only quite a small stimulus. He might suggest that, the pupil should let him turn his head by rotating it to one side. Many people—most people, in fact—when such a movement is suggested, will not be able to let the teacher carry out the movement for them, but will start to do it themselves. The pupil will therefore be told not to do the movement, but, in Alexander phraseology, to "inhibit". In terms of our Input-Throughput-Output diagram (page 161) when a stimulus (the Input) is received, it is necessary not to respond with an immediate muscular Output, but instead to enage in Throughput—that is to say, the projection to oneself of the words, "Head forward and out". Whilst these words are projected (and provided the pupil will "inhibit" the doing of the movement which the teacher says he is going to carry out) it becomes possible to detect just when tension is created in the neck and around the head and throat.

At some point in this procedure, the teacher will have emphasised the importance of the direction "NECK FREE", so that the pupil will come to associate with the direction "NECK FREE, HEAD FORWARD AND OUT", not only a spatial positioning of the skull in relationship to the chest and hump, but also a releasing of neck and throat tensions which have become apparent.

When the teacher adjusts the head, he is able to release quite a lot of tension in the neck and hump: but the pupil will also notice that some slight force is being exerted on the back of the chest and the lower back. When this occurs, the teacher will give the verbal direction "BACK LENGTHEN AND WIDEN", whilst in various and devious ways he achieves such a co-ordination. In lengthening the back, he will be at pains not to produce an undue arch in the middle of the back, and, indeed, by adjustment of the chest and pelvis, he will see to it that the lower back becomes almost completely flat on the couch. As he continues with this procedure, he will insist that the new verbal direction should be in the right order. If, as he adjusts the lower back, the pupil should stiffen his neck, the teacher will insist that the pupil should emphasise to himself the order, "Head forward and out", until it is clearly perceived, and then add on to it the direction to the back to "lengthen and widen". With his hands he will repeat the muscular experiences in the right order—a task which sounds complex when written down in this way, but fairly obvious in practice when undertaken by a skilled instructor.

There are many more such "orders" to be added to the new body-grammar. The teacher may now say that he wishes to move one of the legs so that the hip flexes and the knee points towards the ceiling. He will again insist on the pupil "inhibiting" any doing of the movement, and he will ask him to continue to project the orders to the neck, head and back, and to add on to them such an order as "KNEE OUT OF THE HIP" or perhaps "HIP FREE", or "KNEE TO THE CEILING". (There is no right or wrong about the actual verbal phrasing of such orders, but merely a sequence in which it is best to give attention to the body.)* [45, 46, 47]

Teachers may vary as to the point at which they will tackle the shoulders and arms. I personally think they should very soon be linked up with the instruction to free the neck, since many neck and chest tensions will not release until the whole shoulder girdle—comprising the shoulder-blade and collar-bone—is adjusted. I usually suggest the order "SHOULDER RELEASE AND WIDEN", and since this is always a dramatic adjustment at first, I find it helpful to do this in the first session, since a most obvious feeling of bodily re-adjustment usually takes place and even the dimmest of pupils (kinaesthetically, that is to say) will usually notice when one shoulder now lies two inches wider and lower than the one on the other side. And since the trapezius muscle goes from the back of the head, down the side of the neck, into the shoulder girdle, it is possible to give the subject an easy and obvious demonstration that, unless he thinks of his head going "forward-and-out" as his shoulder is adjusted, his head will simply pull down with the shoulder because of the trapezius contraction. Most people, during this procedure will realise that by giving their attention (in the form of the verbal "orders") they can maintain an improved head-neck-shoulder balance.

But again, the varieties of shoulder-use are considerable: all that can be looked for is a gradual transition to an anatomical norm. And it is explained again and again to the patient that what he is learning is a neutral "resting position" of balance of the various parts of his body—rather like the "neutral" of a car to which one can return after one has been in gear. And it is stressed throughout that for any given position or activity there is a due amount of muscle tension needed. More muscle work will be needed for sitting than for lying: more work for lifting a hammer than a tooth-brush: but that the increase in muscular activity should be undertaken on a general and not a local basis.

* Professor G. E. Coghill whose *Anatomy and the Problem of Behaviour* was considered a classic in its time by such people as Le Gros Clark and J. Z. Young explained that the correct sequence was Cephalo-caudad, from head-to-tail. His book, along with Judson Herrick's biography *G. E. Coghill: Naturalist and Philosopher*, is useful reading since he gave whole-hearted backing to Alexander. I touched on Coghill's account of sequential ordering in an article in *The Lancet*, entitled "Psychosomatic problems in postural re-education".

Increased forearm work should not be produced in a way which involves a hunching of the shoulders and a tensing of one side of the neck. Lack of interference with Alexander's "Primary Control" of the head and neck is taken to be of prime importance at rest and during activity.

It is unlikely that, in the initial session or two, much more can be done than to familiarise the patient with the words of his new body-grammar, and to get him used to being handled in a certain way and a certain sequence: but the impact of this whole procedure is considerable and most people will realise that they are being asked to set about things in a way which is quite novel and individual to themselves. I have had quite young children describing to me—in their later years—the immense impression which such "lessons" had upon them, and the feeling of security and individual attention which they remember.

It is indeed on this fundamental basis of practical attention and instruction that the patient/pupil will eventually be prepared to let go more deep-seated and unconscious tension patterns. It can be frightening to surrender our familiar sense of balance to someone else, and most of us—as I have noted in Chapter 7—have a multitude of tricks of muscle-armouring by which we defend ourselves from contact which we fear to be harmful. The release of tension may lead to anxiety—or tears, or laughter or anger: anything to distract one's attention away from the unfamiliar experiences of releasing tension. This goes for seemingly quite normal people, as well as for those who are in a state of apprehension or distress.

A well-constructed Alexander lesson will not have the effect of making the subject feel that he is being "got at", but that both he and the teacher are learning together how to sort out the mis-use problems. It is not that there is one master-key—Alexander's Primary Control—which will unlock the prison, but rather that there is a key-blank, out of which, by using the Alexander Principle, an individual key can be constructed: a key which can be used to unlock our unnecessary defences and to open up a way of dealing with future buffetings. Life will always buffet us, if we are to live fully.

Sitting and Standing

Many Alexander teachers prefer not to begin instruction

lying down, but in the more active situations of sitting and standing. The subject may be asked to take up a standing position, and it will be emphasised that he is not to "do" the new body-pattern, but simply to project it to himself. He must never "do" the orders: just think them. It will be explained that such "ordering" is always a "pre-" activity: that when we receive an outside stimulus (or an inside one from some idea which has sprung up or which we have cooked up) we are all of us liable to react with preparatory tension: that we are liable to get "set" in preparation for what we are going to do, and that such an anticipatory pre-set usually triggers us off into far too much effort when we initiate a movement.*

When the subject is standing quite still, the teacher will obtain as good a pre-set as he can by his gentle manual adjustment, and he will then ask the subject to continuously project the new body-pattern as the knees are bent forward to sit down. This is an unfamiliar experience at first, in which the usual balance may seem to be alarmingly upset. In time, however, it becomes familiar, and in this specific situation of sitting-down and standing-up, it is possible for much insight to be gained into faulty tensional habits and into the maintenance of the new pattern under stress. Likewise, when in the sitting position, small movements of the trunk backward and forward from the hip-joint (whilst maintaining the improved direction of head, neck and back) can be used to teach the subject how to attend to his USE. It should be explained that in such activities, the subject is not learning a correct way to sit and stand but how to attend to his USE so as to prevent unnecessary tension. Such learning carries over into other situations.

Alexander wrote frequently of a "position of mechanical advantage". An easy way to achieve this is to slide the back down the wall, as described in Chapter 9, at the same time putting the knees apart and flexing the pelvis so that the whole

* A body sliding along a surface is restricted by *friction*: but to start moving along a surface, it is necessary to overcome what engineers call "stiction": we have to get a car moving by using low gears, to overcome the initial inertia. In much the same way, our joints exhibit this property of "stiction". They tend to get relatively stuck when held still and fixed for a time, and an extra effort may be needed to initiate movement. The moral of this is not to allow ourselves to become fixed when at rest, and to let out the clutch slowly when we get up out of a chair or any other movement involving change of position.

back is flat against the wall. If the entire spine from head to pelvis is now kept lengthening, it can be inclined forward from the wall, with the buttocks still touching the wall (fig. 31). In this position the directions are "Head forward and up; back lengthen and widen; knees forward and apart", and in addition, the neck can be directed to lengthen "Up-and-back" and the lumbar spine likewise up-and-back (as in fig. 5b). Plate 13a shows the same manœuvre without a wall.

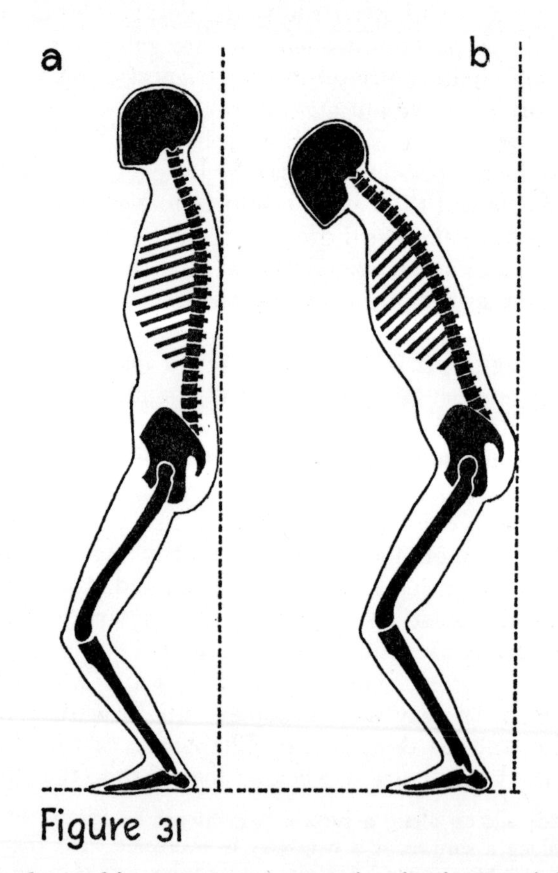

Figure 31

Throughout this manœuvre, attention is given to inhibiting and projecting the new body construct, which includes orders to the knees and ankles and elbows and hands.

Such detailed attention leads to a heightened awareness of the co-ordination of the whole body, and most subjects will experience an exhilarating feeling of lightness and "up-ness" in their bodies as they begin to engage their minds in such a manner.*[48] Such an experience, although unfamiliar, will often convince people that they are engaging in something of great value.

The Alexander Teacher

Alexander teachers, naturally, vary in their approaches, but unlike other teaching situations, they all have one thing in common: they will have to obey their own educational demands if they are to influence their pupils. There is no trick, however clever, by which this fact can be evaded. It is for this reason that the training of Alexander teachers has always insisted on first improving their own USE to a point at which they not only have a high standard but one which they can sustain when they are under full teaching pressure. This does not mean that, in order to teach well, it is necessary for the teacher to be in some ineffable state of "directive lightness". At the other end of the scale, the teaching can never become automatic and rule of thumb, since each pupil presents his own personal problems and difficulties.

A detached form of teaching which relies on a pedagogic, professorial, didactic attitude, is simply not possible. It is intensely boring if during a lesson the teacher gives the impression that he is seeking to expound a thesis or to prove a case, and makes the pupil feel he is being reproved if his attention wanders or if for a moment he disagrees.

But the pupil also must realise that the learning process involves a most detailed attention on his part: that his organism is being "re-calibrated", and brought back to true: that ideally, he will eventually need to give attention to his USE all day long, so that not only will he become freer whilst interacting with the outside world, but will know how to return to a balanced state of rest after such interaction. And both he and his teacher can never forget what a tall order it is to ask him to disobey habit.

It follows that, during a training programme, various emotions may be "transferred" to the teacher. There is often an initial "honeymoon" period of great pleasure in having discovered not only something which explains previous troubles, but which also offers a solution. Some people continue their happy honeymoon into a long and successful "marriage" with

* Many mystical experiences include an "UP" component (*Ecstasy*, Marghanita Laski, Cresset Press, 1961).

the new idea and practices. But for many—and probably most people—there are bound to be "break-down" periods which lead to a certain amount of depression and to a feeling of dissatisfaction that a given Alexander teacher is not being clear or helpful enough. Also, to see oneself as one really is must usually involve some cutting down on, or alteration in, outside commitments, at least for a time. Few people can expect to make such big changes in themselves without experiencing periods of anxiety and depression, and they may find that they are not able to meet the new need and challenge of the Alexander Principle without abandoning, at least for a time, the effort to pursue some other relationship. Added to this, most people feel slightly stupid with their new USE at first, because the new USE-components have not yet been incorporated into familiar sequential rhythms. Ease comes from a feeling of on-going sequence—we can all remember our difficulties in first learning to drive or ski. The new Alexander-USE has no clear place at first in our sequential organisation. When however, the new patterns become incorporated into newly-timed sequences, we get a feeling of ease and a motivation to use the new patterns—an ease which at first can only be experienced when the teacher is giving us help. The enjoyable feeling of "lightness" which comes after working with a good teacher is evanescent at first: and if there is not to be an undue dependence on the teacher who can give us this "lightness"—and with such dependence, all the concomitant problems of the transference situation—then the pupil must learn to work independently, on his own.

At present the majority of Alexander pupils are adults and adolescents who have found their way to a teacher, either through the guidance of their doctors or educational friends, or simply by word of mouth. They form an immensely variegated cross-section of the population, and their problems are equally variegated. There can be no one correct way to learn the Principle—each will find his own way on the basis of his own education and Alexander instruction. I have found in the past that the question and answer approach is a useful way of dealing with so many different attitudes, and in the next chapter this method will be used.

Obviously most of the people who read this chapter will not have had personal Alexander instruction but this is not essential for understanding it.

Chapter 11

LEARNING THE PRINCIPLE

THIS CHAPTER TAKES the form of a series of questions by a pupil or prospective pupil, and answers from an Alexander teacher.

* * *

Just what sort of work am I expected to do on my own in order to achieve a balanced regulation of my body?

The Alexander Technique aims to teach a pupil to associate a new sequence of thought with a new manner of using the body. You may already have found that when running over the new sequence of thought you can become aware of much unnecessary tension at rest, and unnecessary pressure during movement.

That sounds fine, but under the almost incessant stimulation of my busy life, I don't seem to have much time or inclination to think about this sort of thing. I realise that I need to alter my old habits but just where am I to begin?

The situations which, in the majority of people, produce muscle tension, and take them away from a balanced resting state are:

a. Talking to other people, especially to people you know well, particularly when your job or life situation requires you to establish rapport with them.
b. The stimulus of handling familiar objects around you—the toothbrush, door-handle, articles of clothing, typewriter, gear-lever, piano, food, and indeed the whole range of biological activities from swallowing to excreting.
c. The commitments in which you are already engaged, e.g., the tennis player in his team, the office boy who has to be obsequious, the business executive who feels he has to adopt a forceful attitude, the dancer or physical educationist who has to make movements which involve maldistributed tension, the teacher who has to fit her teaching to a curriculum and get things done in a certain time.

d. The emotional gusts: waves of irritation, fear, sexual excitement, crying, and sentimental eye moistening, depression, suspicion, excitability.

e. Obsessional repetitive mental states: day dreaming, recurrent tunes in the head, self conversation, etc. Such states occur as a background to tension which is already present, and the attempt to control them may lead to further tension.

f. Excessive desire for the smoke, the drink, the chocolate, etc.: it is not suggested that such things are harmful in moderation, but many people find themselves in a state of tension and conflict for which these are a poor form of palliation.

g. Fatigue: after making excessive tension there is a temptation to collapse and "slump". It is a mistake to slump in a chair—far better to lie horizontally if one is tired.

h. The feeling of unfamiliarity with the new conditions; feeling, for example, "stuck up" or different when employing the new body-construct.

i. Rush: the need (or imagined need) to get things done quickly.

j. Frustration: civilisation usually implies delay between desire and satisfaction, and under these conditions, tension may mount up unless one consciously regulates it.

But how am I expected to alter these habits?

In your lessons so far, you will have realised the importance of inhibition—"stopping off" an immediate reaction to the stimuli which your teacher has given you. You will have found that, unless you "inhibit" the stimulus, say, of sitting down or standing up, you react in the tense and distorted way which it is necessary to change: but you will have found that, by "inhibiting" and employing the new form of "direction", you can usually prevent this happening. In the same way it should now be possible for you to check your immediate reaction to many of the situations outlined above; and even if you find it impossible to maintain the improved tension-balance, you are at any rate aware that it is undesirable, and are able to return to a better resting equilibrium when the immediate stress is over.

Am I expected to think of this all day long?

In the beginning stages it is most unlikely that you will be able to work at it all day long. Start with the simplest activities, and set aside special periods when you work at this and this only. There are, in addition, very many hours during the day when one must perforce keep still in one place, e.g., waiting for a bus, or many other times when overt activity is prevented or frustrated.

How, then, am I to work in these special periods?

You will probably have learned from your lessons that "realisation" is more important than "trying". Intense concentration for only short periods of time will not produce permanent change in your manner-of-use, but it will begin to build up and to carry over continually into your everyday life. It is a good idea to make a habit of working for short periods in the following way:

i Find somewhere where you are not likely to be disturbed, and where you can lie down if necessary—the floor with a book under your head will do.

ii Lying down with a book under your head and with your knees pointing to the ceiling, decide consciously to keep quite still, i.e., don't wriggle or scratch, and don't follow irrelevant patterns of thought. Give to yourself the following verbal directions:

(*a*) "Neck release, head forward and out." As you give this sequence of thought, at first you will not "realise" what the direction means but as you continue, you will begin to associate it with an awareness of your neck and your head. If you are out of touch with yourself, you will begin to regain the degree of awareness which you may previously have had, either when working well on yourself or during a lesson. By "awareness" is meant what should be a normal sense of "being-in-yourself", as opposed to the state of mind-body split which is so often present in adults, if not in children.

(*b*) While preserving, by direction, this awareness of the head and neck—an awareness in which your verbal order will be part-and-parcel of the actual perception as the organising component of it—add on the verbal direction to "lengthen and widen your back". Your lessons will already have made you familiar with the meaning of this phrase, although it is likely that fresh meaning and fresh simplification will accrue

as you run over the sequence to yourself. For example, the whole of the back may be realised as lengthening in one unit instead of thinking of the upper back as separate from the lower back: or perhaps the "widening" of the back may suddenly be noticed to include a releasing of the shoulder blades and arm-pit. At this point, your interest in the new realisation may be found to have caused you to "lose" the head direction, and it will be necessary to re-inforce this before returning to the lengthening and widening direction.
(*c*) This process of adding together the direction to the head and the direction to the back may take several minutes or even longer: indeed, if it seems to take less time, you will almost certainly have been making a muscular change by direct movement, instead of sticking to realising the meaning of the orders. Remember that we do not move our bodies in the same way as we pick up an external object—a brush or a pen or a pail. To "move" our forearm is not the same as to move a spoon. Moving ourselves and bits of ourselves—as opposed to moving external objects—is always a question of *allowing* movement to take place, rather than of picking-up and putting somewhere else. Allowing the movement, say, of an arm, should involve a total general awareness of the body in which the active process involved in the arm movement is small compared with the active process of awareness which is going on in the whole of the body all of the time. Similarly, a movement of standing up after sitting down—a movement which mainly involves a leg adjustment—does not require only the leg activity, but primarily a maintenance of the awareness of the rest of the body whilst allowing the necessary leg movement to take place.

When I was thinking about it yesterday I noticed certain muscular changes taking place. You have told me I should be wary of trusting my feeling, and also that I should not "do". Was I "doing" this muscular change I felt: in fact was I working the wrong way?

If you were "ordering", it is possible that changes did take place. Most people, however, when they notice such things beginning to happen, stop "ordering" and get interested in helping the changes by "doing". When you notice something beginning to happen, it is more important than ever to "think" and not to "do". It will be found, however, that if you move after having "ordered" for quite a long time, muscular over-tension will release as you move, provided that you continue ordering as you move. In other words, the directions which you have learned to give to yourself affect the "pre-set" (the preparatory tension) in your muscles.

Does it help if I visualise what you are teaching me?
No. The kinaesthetic (muscular) sense is separate from the visual sense. Visual imagery, which is almost certainly associated with old, wrong muscular sensations, is liable to lead to confusion between these old muscular sensations and the new ones you have to learn. It is far safer to use a brand new symbolism to link up with the improved new kinaesthetic sense —a new "body construct" made up of the body-grammar which you have learned.

Could you explain what you mean by "body-construct" more clearly?
In the past there has been many terms suggested for the mental model which we have of ourselves. In 1911, Henry Head used the term "Postural model" and wrote: "by means of a perpetual alteration in posture, we are always building up a postural model of ourselves which constantly changes". He also spoke of "Body-memory" which modifies our perceptions at an unconscious level, so that our spatial consciousness is always bound to be influenced by what has happened before.

The term "body-image" was used by Schilder for the visual, mental and memory images which we may have of our bodies. Macdonald Critchley used the term in a similar way, and for many years neuro-physiologists have sketched out a "homunculus"—a little-man image in the brain cortex—which is taken to represent various areas of the body.

The term "Body-concept" has been used for when people think that they are too tall or too short or that their breasts are too large or their bottoms too big.

Other writers have spoken of the "body percept", by which they mean the momentary perceptions we may have of our body at any given time, irrespective of the construction which we may put upon perceptions, or of the mental factors which led us to obtain that particular perception.

For many reasons I prefer the term "body construct",[49] a term which implies not only the way we "construe" things but also the way we "construct" our responses and organise outside things so that they will appear the way we want them to be. Such a "body construct" produces (and is based on) our habitual USE of our bodies, and it forms the background to our perceptions.

It is a common experience that once we are set and readied for a given course of action, it is impossible to think of doing anything else. Our preparatory "set" affects our observations so that we see everything in conformity to it. We will all of us have had the experience, when waiting to carry out some activity, of being triggered off into action by some quite inappropriate stimulus, some accidental resemblance to the configuration which we are waiting for, or, when we are very strung up, something which bears no resemblance to it whatever. If we are to check such reactions and only release our response when the time is right, we need to have control over our preparatory state. Our subjective experience of this preparatory state is what I understand by our "body construct".

Does this mean that the new "body construct" will eventually become unconscious?

No. A human being is not a one-way system, unconsciously reacting on a stimulus/response basis according to previous conditioning. We ourselves search out from our surroundings the stimuli which we prefer to perceive. We don't sit around waiting to salivate when a dinner gong may ring: we are constantly concerned with the "organisation of preferred perceptions". Such preferences make up the "body construct" and are embedded in our habitual resting state. The way we use our body is the way we perceive and construe our world. The organisation of our preferred perception has its basis in the use of our muscles whether they be used in posture, movement or communication.

Such preferences are usually at an unchosen, unconscious level: our characteristic preferences are embodied in our muscular tension patterns. The Alexander Principle suggests that our "body construct" must be used consciously and that this will involve us in "inhibiting" our habitual responses.

The repeated use of such "inhibiting" and projecting a conscious body construct would be an impossibility in everyday life, unless some degree of amalgamative learning had taken place. At first, the new details of muscular USE may have to be "thought-out": but in time, this becomes a state of "thoughtful movement", rather than of "thought-out action".

To understand how this happens, let us consider the learning of some special skill like learning to drive a car. The way in which driving a car becomes "second nature" is familiar to many of us—the painstaking need to at first remember a right order for doing each movement, until this grows into a familiar routine which does not have to be thought-out each time. And not only a familiar routine: the outline of the car will eventually become an extension of our personal body boundaries. The body-construct has matured, and we can now release the "thinking-out" part of our brain for the moment-to-moment control of the car in traffic. There will no longer be the need to "think-out" each detail: but there will still be the need for "thoughtful-movement"—thought which concerns itself with the general USE, as well as with arm and feet movements.

A new "body construct", once learned, can be used to put oneself into a state of "thoughtful movement": but there will always have to be a decision to switch it on. It has to be made, and made afresh, according to the circumstances. It may not be immediately accessible to us at a conscious level, as part of our ordinary awareness, but has to be "directed".

Will working in this way help me to cope better with things?

Yes, but only if you are prepared to work hard at it and show considerable initiative. There are bound to be discouragements which tempt you back to your familiar pattern, and there may be a period in which you manage certain aspects of your life less effectively because of the need to develop the new pattern: just as a tennis player, cricketer or golfer may have to re-learn his technique in order to get the "bugs" out of it, and may for a time be less efficient. Indeed, working in this new way may lead at first to discomfort-plus-desire-for-change, rather than comfort-plus-coping-better-with-the-present. Nevertheless, conditions of physical pain may clear up fairly soon, although a temporary situation may well arise in which the pain clears up provided one works in this new unfamiliar manner, but the pain returns when one adopts a familiar "comfortable" pattern, e.g., slumping, crossing the knees, or reacting in an over-excited manner.

Can you give me some advice about my breathing?

It is easier to detect the more subtle faults of breathing when lying down, but first take a look at yourself standing. Preferably remove all your clothing and stand in front of a long mirror, arms hanging down. Are the tips of your fingers at the same level?

If one hand is lower than the other, look at your shoulders. One will be lower than the other. Look at the line of your neck. It will be longer on the dropped side than on the other side.

Now imagine a perpendicular dropping from the outer edge of the lower shoulder towards the floor (fig. 32a). It will probably fall through your thigh. Now look at the perpendicular from the other shoulder. It will probably miss the thigh by about an inch. Why?

Look at your whole chest cage. It is because your chest is displaced sideways (see plate 19) that your hand and shoulder are lower. This sideways displacement will affect breathing on the dropped side, and the breathing may also be further affected by a rotation of the chest backward on one side.

Now look at yourself sideways on (fig. 32b). Observe the base of the neck at the back. If the neck vertebrae are dropping too far forward and the spine at the back of the chest is bent forward, this produces an excessive hump. The forward dropping of the neck will exert pressure on the trachea. It is necessary to carry the head on the neck in such a way that the forward curve of the neck is corrected.

Now look at your lower back. If your back is arched (figs. 32b and 32c) and your abdomen is protruding, your chest cage in front will, usually, be pushed forward, and in most people the angle between the ribs in front will be narrowed. It is difficult to get full breathing movement unless the arching of the back is corrected. This must be corrected in such a way as not to accentuate the hump at the base of the neck.

Lie on your back on a hard surface with your knees pointing upward and a support under your head. Observe the bend of your elbows. The funny bone should be turned outward from the body and the inside of the elbow towards the body, so as to widen the armpit and upper arm away from the sides of the chest. If, in this position, your forearms and hands will not lie flat on the surface (fig. 32c), your shoulder girdles are too tense and need to be released from the base of the neck.

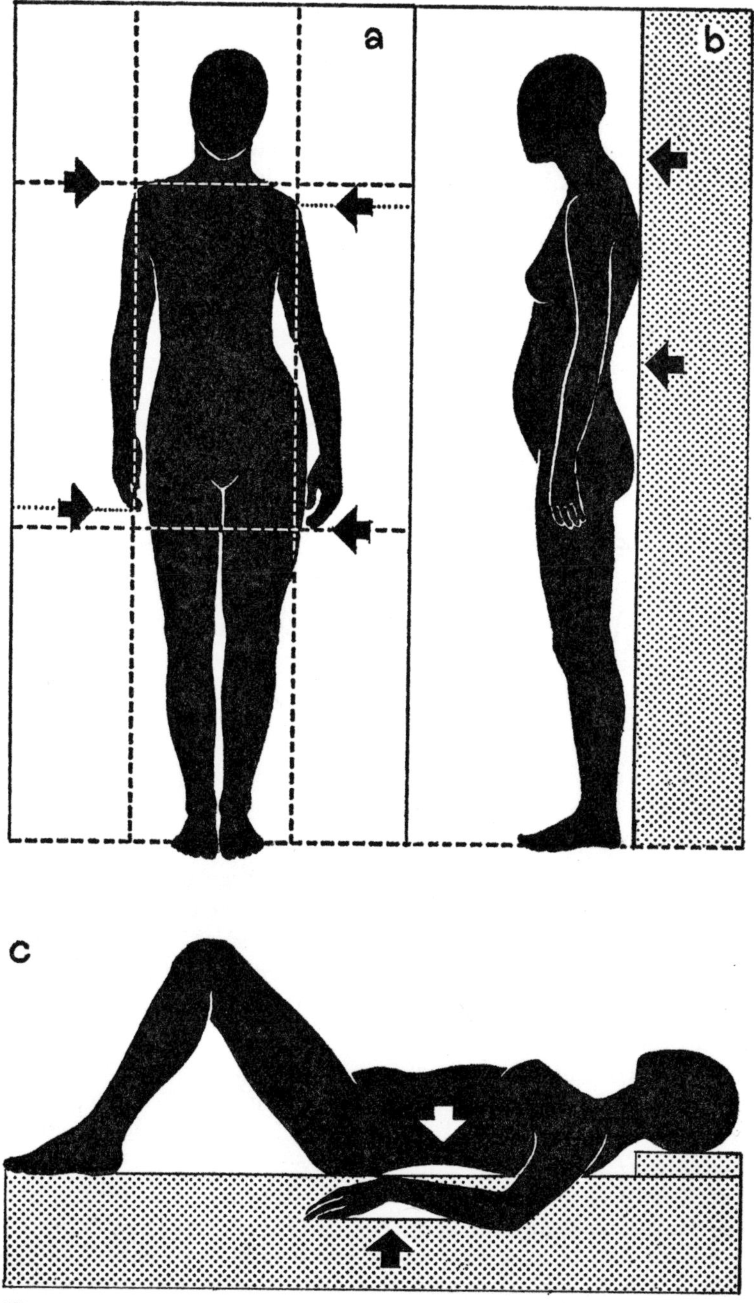

Figure 32

Figure 32a shows dropped left shoulder and arm, thorax displaced to right, pelvis raised on left. 32b shows neck dropped forward and back unduly arched. 32c shows back wrongly arched off surface and forearm not lying flat.

Are you breathing? Many people, when studying or concentrating, hold their breath for long periods. Don't. At rest breathe at least ten to twelve times a minute.

Are you moving your chest and abdomen in front when you start to breathe in? Breathing in is a *back* activity. If you start breathing in by raising your upper chest in front, it is like trying to open an umbrella by pulling on the cover from the outside at the top. It can be done, but it is inefficient.

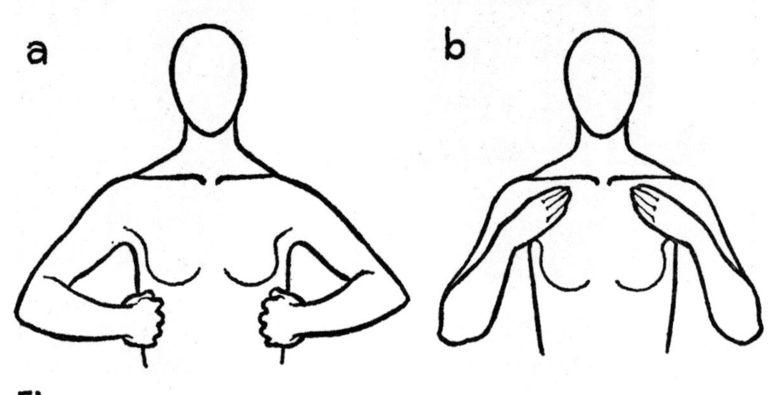

Figure 33

Place the backs of your hands against the sides of your chest (fig. 33a). Imagine the gills of a fish half-way down your back on each side. Breathing in should start there, and the ribs should move out sideways against your hands. If your chest cage is displaced sideways, one side will move more than the other. Is the bottom of the rib cage nearer to your pelvis on one side than the other?

Place your hands on the upper chest (fig. 33b), just below the collar bones on each side, and almost touching the breast bone. When you start to breathe out, there should be a slight release of tension as the upper chest and breast bone drop. If you are very tense, a sigh will give you the feeling of the upper chest releasing. Don't raise the chest when you start to breathe in.

Breathing out, at rest, should last at least twice as long as breathing in. As you finish breathing out, you will feel your stomach muscles contract slightly. In order to get the next breath into your back, you will first have to release this stomach contraction. Many breathing difficulties come from keeping the upper chest and abdominal muscles too tense in front even at rest.

Look at your nostrils. Is one less dilated than the other? If you touch it, it will spring out slightly. Where is your tongue? At rest it should not press on the roof of the mouth but should lie flat on the floor of the mouth. The dilated nostril and flattened tongue will give you a better airway.

Think about your throat. When a baby screams, or when you attempt to defaecate, you will notice a tightening in the throat. Some people, when they breathe in, tighten in this region and do not release it completely when breathing out. This leads to fixation of the upper chest—an attitude often betokening fear of aggression. This tension can be relaxed, on breathing out, by releasing the throat and dropping the upper chest very slightly.

Look at your shoulders. The shoulder-blades should never be pulled together at the back—there has been much faulty teaching about this in the past—but they should lie flat on the chest cage (plates 12d and 26).

I cannot help feeling, from what you have written that you think that no one, in the long run, can be really happy or healthy unless they use the Alexander Principle. Most medical and educational innovations tend to be "over-sold" when they are first put forward. Don't you think that you have spoiled your case by saying such things as, "This seems to me to be the single most important problem which medicine now has to deal with."

The objection would certainly be valid if I had suggested that the Alexander Principle was a "cure-all". But this isn't the point which I am making. The recognition of the widespread presence of bacteria was not a "cure-all"—it simply pointed the way in which research had to be directed. It is the same with the mis-use: its manifestations need to be tabulated as carefully as were the types of bacteria. The recognition of the widespread presence of mis-use opens up a whole new field of Preventive Medicine, whilst at the same time indicating certain medical conditions—notably the rheumatic and mental disorders—in which immediate application can be made.

It would indeed be unhelpful to hold out something which in fact cannot be provided: or to promise great prospects to people who theoretically might be capable of working in this way, but who, when it comes down to it, will be quite incapable of it. There is so much that is clear and valuable in this approach that it would be silly to claim too much. The influence of USE on mental and physical functioning (and, therefore, on "health and happiness") is quite clear: what we can never be so clear about is the extent of the trainability of any given person, and their willingness to use what we can teach them. There is, however, in the majority of people an area, however small, in which their USE can be improved, with consequent improvement in their functioning (though not the "cure" of their "disease").

In the educational sphere, there is no question of over-selling the Alexander Principle. The training of teachers of it has been recognised by the Inner London Education Authority and selected students at the Alexander Institute are eligible for training grants. The problem is to make a sufficient number of well-trained teachers available to satisfy the demand.

Chapter 12

APPLYING THE PRINCIPLE

Why are we so fagged, so fashed, so cogged, so cumbered,
When the thing we freely forfeit is kept with fonder a care?

HOPKINS (*The Golden Echo*)

MOST POLEMICAL BOOKS—and this book is intended, at least in part, to be polemical—are strong on the diagnosis of predicament, but not so strong on solution. The Alexander Principle poses a new sort of predicament: it proposes an inclusive hypothesis-bag into which to stuff several of our present-day predicaments. It says that it is better to live by a difficult principle which unifies, than by a series of lower-order rules which constantly come into conflict with each other.

Like all hypotheses it has its limitations and it needs testing and refining: but a sufficient number of people have lived successfully by it for a sufficient number of years to indicate that it is a viable proposition.

The Mature Rebellion

There comes a time in most people's lives when they feel the need to take a hard look at their pattern of living. Adolescent rebellion is, in the main, a rebellion against external life-forces —parents, teachers, employers and politicians who are blamed for the difficulties and inequities which become only too clear once childhood is over. The vast impersonal world moves implacably on, and the desire for someone at least to take some notice of us, plus the desire to be rid of adult deceit and the desire not to be "grey", leads us, in adolescence, to upheaval and disruption. The young are living and growing and life is to be sensed and enjoyed. They are no longer willing to put up with the barrack-square attitude "You'll bloody well do as you're bloody well *told*, same as I do" (a choice morsel which floated through the windows of my Medical Inspection room during the war). They wish to be rid of the deadening junk of the past, the customs which enslave them, and the false respect which stupid people seem to demand from them.

But one can't go on blaming the older people and the outside world for ever. A more mature rebellion stops saying "The world is making me grey: they hate me without cause", and starts to say instead, "The buck stops here with me: just exactly what is the matter with me: why don't I do as well as I know I can, why does my back ache, why is my sex a muddle, why am I ten different people: how can I begin to take responsibility for my own problems?"

This *internal* rebellion is a request for valid Principles, not for a revolution. The rebel is not necessarily a revolutionary. The rebel is determinedly unsubdued by the pressure of external authority, but this does not lead him to a revolutionary world of terror. The rebel has no wish to exploit anyone, but simply a desire to escape from the deadening octopus of mediocrity which tells him to be as dull as everyone else.

If he is to escape from his dullness he needs a plan of action: he needs principles with which to escape the "system".

There is no dearth of helpful suggestions. The priest, the psychiatrist, the poet, the novelist have much to say about the inner-life: the sociologist and doctor can recommend job-change, wife-change, recreation-change, diet-change, medicine and drugs. But individual men and women, in the main, still lead lives of quiet desperation, in spite of 2,000 years of religion, art and science. They have periods of great humour and joy, excitement and success; but, by and large, far too many people's plans go awry, their health and happiness deteriorate, their drive peters out in frustrated boredom. They live two lives: a Personal Life, based on the home and family, where at least they feel real, if miserable; and the Working Life, with its values of money-seeking and social status, where they experience a quite different reality: and if they are un-unemployed, doing a mundane job for lack of anything better, they become un-real people, in bondage to a life of tedium.

This is the lot of many apparently "normal" people. But even "normal" people will become disorganised and eventually break down when they are faced with seemingly insoluble situations. And when they become depressed and "got-down" by the complexity of it all, they begin to accentuate their latent habits of mis-USE.

The Core-Structure

I have used the phrase "core-structure" to indicate the development of a personal standard of good USE which can stop us from being "got-down" and by means of which we can preserve our onward momentum and vitality even when external stress makes stability difficult: a core-structure which will sustain us also when long periods of necessary learning have not yet given us skill and confidence.

Amidst the give-and-take of personal relationship, with its "posture-swapping" and role-taking, such a core-structure enables due respect to be given to other people's needs and attitudes without us having to copy their mis-USE. It does away with the need for the "conformity-deformities" by which we are expected to show loyalty to our tribe; but it also gives a capacity to reconcile, by giving a capacity to maintain our own stability in the presence of conflict, and in the presence of unpleasant end-gainers. In a world in which we may have little in common with some people, it helps us to find common ground.

Most especially it helps in the moment-to-moment regulation of unwanted emotion. One of my patients expressed the help which she gets from it in the following way.

> Often I get a feeling of being *unsafe*: it is not an easy feeling to describe exactly, but it must date right back to childhood. The feeling is caused by other people and comes on quite suddenly. A look or a word is enough to start the feeling, which hits me in my midriff and it makes me feel knotted up and apprehensive. Since working with you I have discovered that physically I contract the part of my body which lies just below my ribs (this must be the region the ancients called the solar plexus) and it also has the effect of making me arch the middle of my back. This seems to be the way in which I produce feelings of unsafety in myself. If I can catch myself in time before this contraction sets in, the unsafe feeling only lasts for a moment. I have discovered that if I can remember to give my "orders" when I begin to get this feeling, I can stop it getting a grip. It stops me going into this quite dotty feeling of insecurity, in which I feel as if everyone is getting at me in a threatening way. "Ordering" seems to make me able to adjust to the other person, not necessarily very well, but at least not stupidly. I can react from a secure feeling of widening across my back and shoulders, instead of from a panic feeling in my stomach. It also makes me often able to see that no

reaction is called-for, and that if I stay quite calm, with a feeling of being supported in my back, I can cope perfectly well with what is going on. I can see when it is appropriate to react, and when I needn't.

Feelings of safety or lack of security date from our earliest childhood, and are based on our feeling of our USE. A loving serene home atmosphere with plenty of warm physical contact, can certainly help to make a child feel secure. Unfortunately this feeling of security usually also includes a copying of the parents postures and moods and corresponding mis-uses. The feeling of "right" and the feeling of security which we get from this posture may serve us well on the home-ground (and many never leave the home-ground) but it cannot be appropriate for other situations which we must explore if we are to grow. Without a manner of USE which we can make basic to *all* situations, we are liable to feel insecurity and to confine ourselves within the bounds of what we know already—the recipe for a humdrum life.

If parents know something about the principles of good USE, the child's feeling of security can be based on such good USE, and with it a loving, communicative and non-aggressive relationship.

The Reasonable Man

The patient mentioned above, with her feelings of insecurity and stomach-panic, said she could prevent it happening if she "gave her orders". Just what exactly was she talking about?

I described in Chapters 10 and 11 the teaching and learning of a sequence of "orders", which gradually comes to be associated with a better resting-state of balance, and with a core-structure which gives a more secure balance during movement and communication. Alexander considered that by teaching people this ordering-process, he was putting them, as he expressed it, "into communication with their reason", as opposed to the panic and insecurity which may come when we react by instinctive end-gaining.

"Reason" has had a poor press in recent years. The ways and habits of the reasonable man have been at a discount, along with the image of the "good" man. The holocaust of violence which was unleashed in World War II, plus the threat of the nuclear-bomb poised above our heads, has produced an "eat, drink and be irrational" mood—a mood of despair that our religions and our reason should have led us to the abyss.

This has also been the half-century of the "common man": the depressed classes have become more literate and more vocal, but often not sufficiently educated to find and express their own reason. And one curious phenomenon has been a widespread acceptance of the dominant postural mood of the socially-deprived classes—a posture of sullen collapse, plus an aggressive muscular contraction of the shoulders and arms: a posture which has come to be adopted by large numbers of middle-class adolescents as a way of connecting-up with those they sympathise with, and as an escape from the rigid strait-jacket of the army and the genteel.

Caught between a Victorian ethic which believed in reason and self-help, and a Freudian ethic which says that reason and self-help are bound to founder on the shoals of the unconscious, modern man has had few principles which he can use intelligently: and instead he has fallen back on the *ad hoc* satisfaction of his needs, as and when they arise. In this he relies mainly on instinct and on the habits with which he has got by in the past.

Instinct and habit have one thing in common: they mean that we react only to *one* specific aspect of the general situation which confronts us. Reason on the other hand implies that as many as possible of our relevant needs are taken into account before we react: and accordingly it has seemed essential that the rational man should know as much as possible about his "relevant desires and needs". Years and years of school and university education are devoted to learning such relevant reasons, so that, in theory, we can become free to work out suitable alternatives, and to set up our own "oughts" and imperatives.

But the gulf between theory and practice remains to be bridged. Reason, when learned only at this level, has proved inadequate: and this inadequacy has led, in part, to the present discontent, to the demand even for "de-schooling".

And it has led to a growing need for the Alexander Principle. In Huxley's words:[50]

> It is now possible to conceive of a totally new type of education, affecting the entire range of human activity, from the

physiological, through the intellectual, moral and practical, to the spiritual—an education which, by teaching them proper Use, would preserve children and adults from most of the diseases and evil habits that now affect them: an education whose training would provide men and women with the psycho-physical means for behaving rationally.

Reason, as taught in school and in university will always be liable to founder, unless the reasoning faculty can be used in an additional way, TO DIRECT OUR DECISIONS. Alexander—and of course many, many people before him—realised that the reasoning *deciding* part of our consciousness could also be employed to give directions (orders) to ourselves. But, prior to Alexander, such "directing" had never been applied to USE in anything like the detail which he showed to be necessary: indeed, it could not be, because not enough was known about the sort of USE to which such reasoning direction could be applied. Direction-of-USE is the missing tool which Reason requires, if it is not to "founder on the shoals of unconscious habit".

Thinking-in-Activity

However simple something may be in practice, to describe it is not so simple, and it may require familiarity with quite small detail if it is to be understood. "Directing" the manner-of-use is a simple single activity, but the description is complicated by the varieties of human mis-use.

Because of this difficulty, the Alexander Principle has been over-simplified by some people into the need just to sit up straight or to adopt an improved standing-posture: but these postures are only the outward visible signs of an inwardly directed tensional balance. They are an essential long-term part of the reasoning man's equipment—no more and no less: and it is clear that when, say, we are collapsed down in a state of physical depression, it will be hard to get the ideas which will indicate a way out, and even harder to act on them. But there is much more to applying the Principle than sitting up straight.

Let us postulate a perfect environment—a sort of scientific monastery in which the world is not too much with us: a world in which both reason and faith are considered valid: faith, to let go of the known, reason to work for the unknown: a world which seeks to work by principle and not by habit.

This perfect environment is provided to some extent in an Alexander training-group, although of necessity only for a few hours a day, and not far from the customary pressures of money and family and the future. If I first describe such ideal training conditions, in which Alexander-work has been fined down to its most accurate and most intimate, we can then proceed to "trim" a little from this ideal setting, to see what is possible for everyday man in everyday life.

Under such conditions—whether or not a teacher is giving personal instruction at any given moment—there will be a pre-occupation with thought and movement. End-gaining—action which does not pay heed to the manner-of-use—is less likely to occur under such conditions, because there are none of the raucous stimuli of the bustling world to demand instant response. Subjectively, the distinction will begin to emerge between the "content" of the thought, and the actual "function" of thinking (in the sense that a car must be actually functioning before it can go to this place or that): and it will come to be appreciated that there is more to thinking than manipulating ideas; more than doing sums, more than making sex-plans, more than letters and loving and music and how to arrange the furniture. There will be a growing awareness of another type of thinking, a thinking which is part of movement and stillness, belonging with them and underlying them.

It is with this type of thinking that the Alexander "student" will concern himself within his training group. A state of bodily stillness will be sought in which there is a personal organisation of the Use-perception. This personal organisation ("directing") may be found to involve a very slight muscular activity, an adjustment which is present both at rest and during action. Muscular activity, as we have seen, is never still: it may be fined-down and fined-down, but the gradation between stillness and activity is only one of degree. Slight oscillation (Chapter 3) is always present, even in the balanced resting-state.

Under such conditions of directed thinking, the student will become increasingly aware of the muscular matrix of his decision, and of the part he can play in attending to the small shifts of muscle-tension which will accompany both his emotions and his insights. The appreciation of these shifts is as delicate as the finest touch of the violinist, and such directed thinking is at first a tenuous thing—any fatigue or lessening of attention can put an end to it.

We have usually taken it for granted that we can only use our minds in two deliberate ways—content-thinking (i.e., with words, sentences, music, images, etc.,) and behaviour-control. But between content-thinking and overt behaviour there is another sphere of personal life, a vast world of existence to be managed by awareness and attention (although "managed" is too forceful a term for the attentive living which is implied).

This sort of work has been going on for many years in Alexander groups. It does, however, pre-suppose a degree of peace and quiet which may not be easily available, except perhaps in a hospital, a school or a university. For the average patient (or pupil) who studies the Alexander Principle on his own such detailed application to "directing" is difficult in our present economy. With increasing leisure it should be easier, and most of us are in fact far less busy than our "office hours" mighti mply. Certainly in the initial learning-phase, it is essential to have some freedom from intrusion: but even when there are others around, most people already have a way of keeping up their own "silent soliloquy": and this soliloquy could more profitably be used occasionally in Use-direction. We cannot escape from our function of being attentive human beings, except by living in a deadening state of passivity. A critical sifting process is going on, to a greater or lesser degree, in all of our normal perception. Giving "directions" is like setting the focus and speed of a camera. If the focus is wrong, a blurred picture will result, which can be misinterpreted in many ways. Time spent in directing is never wasted. A far more appropriate response is possible if the focus of perception has been sharpened by directing.

The Sense of Reality

Such increased self-awareness is bound to interfere with some of our social shams and to compel us to face up to reality.*[51]

At core, most of us want reality: but when we are in a mis-

* Pasternak: "The great majority of us are required to live a life of constant duplicity. Your health is bound to be affected if, day-after-day, you say the opposite of what you feel, if you grovel before what you dislike." *Dr Zhivago*. Collins, 1958.

used state, we may have to be *taught* reality. In the Alexander teaching-situation, a teacher can give us a glimpse of reality through the teaching-process—there is usually a chink through which a start can be made.

However, at first it may only be a glimpse: much of the early learning stages will be vague and we may have to content ourselves with an intellectual confidence in the diagnosis and in the procedures.

The process of slowly laying foundations may at first give only a sense of "un-realised potential" rather than one of "real actuality". This state of unrealised potential is distasteful to many of us, and we therefore may prefer the actuality of our old mis-use, however un-real it may be. The discarding of this un-real actuality may seem like a loss at first, and loss of anything familiar is sad: we will find that we are profoundly attached to our old mis-uses and the moods they maintain.

We fear new things, because we fear that we may lose our familiar command of the situation. By learning a familiar command of our USE, we can acquire an abiding sense of consistency, and dispense with the need for an external familiarity. There has to be a willingness to accept the unfamiliar—a deeply implicit feeling that life is open to be changed.

Routines of course will always be needed—all sensible life is based on personally-selected routines; but routines can always be changed, whereas habits can too easily outlive their usefulness and lead us into unreality.

The discarding of such unreality must go on. To *be* what we can be, we have to find out what we are. *Being* what we are means being alive, happy and adjusted to our day-to-day life: in other words, being "all-there". This "all-there" sense of reality is easier for children, who live in a restricted environment where their vivid sensibility allows them to "find-out" all the time. Unless we, as adults find out what we are up to with our USE—and continue to find out about our USE as things change—we will have an increasing sense of un-reality, of not being all-there.

To find out what we are is, however, only to find out the ordinariness, not the supernatural. Reality, when we find it, is ordinary and everyday—but an ordinariness in which our heightened senses can delight. It is not a question of going from the prosaic to the "miraculous", but from unreality to reality.

The Alexander Principle proposes that we can achieve a more real state if we will learn to direct a free manner-of-use. Such a new sense of reality will not occur immediately, except in glimpses: but if the desired pattern is constantly held in mind during the long period of learning, it begins to take over and to take its part in the inevitable processes of growth and change.

Such detailed directing is not easy for most people in every-day life: but such detail is not needed all the time. There is a difference between learning and living (although the fullest life will always be one which seeks to learn something from the next thing we do). However, when we are, for a time, away from the actual pressures of living—whether it be because we have broken down mentally or physically, or because we have deliberately put ourselves in a restricted learning environment—living and learning can become synonymous.

Hypnosis and Relaxation

I am often asked why I don't use hypnosis in order to help people to release their unconscious tensions. Recently a young woman suggested to me that since she was prepared to accept that her neck and chest tensions were due to some infantile trauma, might she not be able to recall the original traumatic incident if she was hypnotised? And that if she were subsequently told of this forgotten reason, might she not then more easily release the tension (a combination of pre-hypnosis and hypnosis)?

I made the following points to her:

> The snag with all such approaches is that your chest-armouring is not now one single specific mis-use, although it may have started as a specific mis-use. Over the intervening years, a most elaborate system of compensating mis-uses has been established, and the discovery of one original "trigger" cause, will not now release your other compensations. Indeed your chest tensions could only have become so firmly established because your *general* manner-of-use at the original time was already disturbed. The soil was there already. Most of us have been through many incidents of extreme fear and

stress when young and if our manner-of-use is good we throw them off. The same thing applies to the extremes of stress which we encounter when we are adults.

There is no need to bring in hypnosis—it can only add another factor of unconscious dependence. Instead of searching for the old context, it is more useful to construct a *new* context, in the light of which the old attitude will be seen now to be insignificant. During your Alexander instruction you will have many insights, once you have begun to establish a resting-state of calmer USE. In such a resting-state, almost anything can serve to remind you of forgotten incidents. Seeing an advertisement on the underground, a chance facial expression, a line of poetry or an overheard remark can serve to flood your awareness with a realisation of some forgotten attitude.

This forgotten attitude was previously manifested in your old mis-use pattern. With Alexander instruction, you have for a time lost the old mis-use pattern, but it will have been triggered into reappearance by, say, the underground poster. Against the background of your newly-found awareness of better USE, you are then able to detect the abrupt transition back to your old familiar (but previously un-felt) state of mis-use.

It follows that when you are having Alexander instruction and are thinking a good deal about it, you should observe quite fleeting insights, and endeavour to think them through, both on your own and with your teacher. An insight may be extremely tenuous and may not necessarily be formulated as a clear attitude to this or that person or situation. It may take the form of a sensation of some new release of muscle-tension, or of some new spatial awareness of how, say, your chest joins on to your lower back; or of how the pelvis can move or release in an integrated manner.

When the fog of mis-use begins to clear, the impact is considerable, and there will be a desire to employ the fog-clearing mechanisms whenever the fog has descended again: as descend it will, although it will now descend on someone who has hope and a method for coping with it.

What has been said about hypnosis applies equally to "relaxation" therapy. I read in a recent treatise on relaxation that "The exercises are not difficult—in fact they are very simple." In a sense this is true. They are simple and ineffectual, except at a trivial level. How possibly could tension habits which have been built up over many years be altered by some simple rule-of-thumb procedure? There has been an alarming vogue recently among psychotherapists of trying to get their patients to relax by various forms of body-contact and gentle handling. The therapist has usually picked up these techniques after a very few instructional sessions from this or that pundit. I am not surprised that the whole subject has got a reputation for being dangerous. The technical problems of teaching a balanced resting-state of USE are teasingly difficult: I was myself nearly ten years in training before I reckoned that it would be sensible to work without supervision.

There cannot be a place for such simple relaxation-therapy, at any serious level. It can never be a question of detecting faulty tension patterns once and for all, de-conditioning them by hypnosis or relaxation, and seeing them disappear. It is rather a matter of continually having to notice the tensions, in countless different situations, and gradually finding out the compensatory-tensions, which, like layers of an onion, manifest themselves when succeeding layers have been stripped off. Just how far the stripping should go will depend on time and inclination, and it is essential that the psychotherapist or teacher should know exactly what he is doing and how much can be tolerated.

USE must not be ignored, whatever the psychotherapeutic situation. Modern linguistics has an adage "Don't ask for The Meaning, ask for the Use"—implying that words should not be studied in isolation from the various sentences in which they take part. In the same way, we should say "Don't look for the memory, look for the Use". The grammar of the body has to be studied in action, in living Use, not dormant in dreams or on the dissecting-table. The meaning of a memory belongs with its present-moment Use: when a mis-use is reconciled back into a directive resting-state, the harmful memory has gone.

The unconscious is irrelevant when we learn to live more consciously—a tautology, but the one must imply the other. Body-use *is* the "unconscious" for most people. As USE becomes more conscious, the unconscious habit can lose its grip.

An Alexander Life-style

It will be clear from this chapter that the Alexander Principle can be applied at many levels. Only a few people at present have the time or inclination to embark on a detailed research into it, and indeed much of the work already done has produced information which can be passed on without similar effort. Many people—probably the majority—will be content, and thankful, to use it for the alleviation of ill-health: the body is resilient, and it can get along fairly well without having to be in some ideally-used state of directive detachment all day long, provided that certain clear general principles of good-use are sustained. There is a distinction of degree between the medical and the educational needs: doctors, by and large, tend to see bacteria rather than infinity in a "grain of sand". It is not an ignoble use of the Principle to remove pain and suffering, as some Alexander adherents have seemed to imply.

But few people will, or can, apply the Principle without sorting out at least some of their basic living-habits: and many, in time, come to value it much more for its effect on their general life-style than for its effect on their corns.

A Principle which seeks to change habit at the very moment of reaction is bound to have an impact on anyone who attempts it, and there have been many and varied attempts by different people to give it a style which fitted their own temperament. Huxley's "non-attachment", Dewey's brand of Pragmatism, Bernard Shaw's Life-force, Ludovici's "race of mental giants", the rationalism of a few minor rationalists, the opportunism of a score of European and American "therapists", who thought it was a band-wagon, but found it was a bed of nails: the short-cutters in the world of speech-training and PE, who listened impatiently to all the talk about "end-gaining": and above all, the crackpots, carnivorous, herbivorous, astrologising, worshipping, finding a haven in Alexander's "Nil a me alienum puto, so long as you keep quiet and don't pull your head back." He once said to me, "If there is a crack-pot within 50 miles, he will find his way to me."

It is a miracle that it survived some of this bunch, and since, in the early days, anyone who was dim-witted or cross at his own learning-difficulties could call it a load of rubbish, Alexander was at the mercy of malevolents, since there was no one to contradict them. It was a buyer's market for most of his life, and it is not surprising that he accepted gratefully the plaudits of any minor "authority" who was prepared to say that his work was valuable. And it is not surprising that he was distrustful of much of the human race—"lowly evolved swine" was one of his epithets in later years, and I can personally vouch that it is true that he kept a blunderbuss in his room with which to scare off itinerant musicians and street-criers—a splendid spectacle of Edwardian outrage. And he used it metaphorically on those who would have liked him to be their "little man round the corner who is a *wizard*".

Fortunately there has always been a hard core of common sense at the back of it all. The actual practical teaching procedures have gone on, individually and in groups, in an almost total separation from what anyone might be saying about it (and as they will continue to go on, relatively undisturbed by anything which I am writing about it).

This is not to say that most of Alexander's early supporters were off-the-beam. John Dewey, Aldous Huxley, Raymond Dart, Canon Shirley, Irene Tasker, Lord Lytton, Isobel Cripps, Dr Peter Macdonald, Fred Watts (his publisher)—all of them were major figures who gave the Principle an essential impetus at various times. And, latterly there have been many others who have seen that the Principle should play an integral part in their Institutions, and who have struggled to make a place for it: along with a large number of non-professional people who have seen that it should play an integral part in their personal living.

They have discovered that what Alexander called the "means-whereby" approach, works for them also. That they can tackle most of their problems by working out the "means": that if they will content themselves with a potential which cannot be immediately cashed, then almost any problem can be broken down into the "next thing" to do: and that a combination of tiny grains of sand eventually makes quite a tidy beach.

Alexander

There were, as I have pointed out, two sides to Alexander's character. He himself never quite escaped the need for the "media". He had to get the ball rolling somehow. He had to make enough money to sustain his own scientific life: and since he was originally an actor, he needed applause, and he thought that others would gauge his success by the extent of that applause.

Yet for at least eight hours of each working day, none of this mattered. He would be working, with a detailed intensity, at the moment-to-moment re-education of his pupils and his groups, with humour and intelligence, giving all the time in the world to the needs of the particular person in front of him. This seemed to me to be genius: no one had ever done just this type of work before.

In common with most doctors, my life has brought me into contact with many very intelligent people—many of them people of the highest talent. For what it is worth, I must place on record that I found in Alexander an imaginative genius and an adherence to scientific method which I have not seen outmatched by anyone. I think he transformed the human condition, although as yet on a tiny scale.

CONCLUSION

> Myself when young did eagerly frequent
> Doctor and Saint and heard great Argument
> About it and about:
> But evermore came out
> By that same door wherein I went.
>
> EDWARD FITZGERALD

ALEXANDER WAS NO saint, but he saw his Principle as a major hope for most of us. Many people have found that his Principle has opened a new door for them: not necessarily the door to wealth and success, but an escape-hatch and a possible way through to what might seem an impossible personal evolution.

Fagged, fashed, cogged, cumbered, stuck in a dreary set of commitments and habits—some evolution must be attempted if a life is to be made with style and with pleasure. A very few people have over the ages chosen to explore something quite different. We have had a half-century of the common man, of the individual doing-his-own-thing. Perhaps we can now have a next half-century of the extra-vidual, the man who not only has freed himself from the compulsions of society, but who has a way of freeing himself from his own compulsions: a way which by the cultivation of an attentive resting-state of USE can move him towards the life of the "whole man". Such a man, *capax universi*, will find a world of infinite charm and variety in that which Alexander has to offer.

APPENDICES

A SHORT BIOGRAPHY OF ALEXANDER

Some of this material is taken from an unpublished autobiographical fragment

FREDERICK MATTHIAS ALEXANDER was born in 1869 at Wynyard on the north-west coast of Tasmania, the oldest son of John and Betsy Alexander. His grandfather, Matthias Alexander, owned a large property which included Table Cape, between the sea and River Inglis. He spent his childhood in country pursuits, and in particular the training and management of horses.

At the age of 17, he started working in the Mount Bischoff Tin Mining Company—"with deep regret I finished the way of life I had enjoyed up to that time, the outdoor experience in the fields, the sea and the beaches". Whilst at Mount Bischoff his main interests were "teaching myself to play the violin and taking part in amateur dramatic performances".

At the age of 20, he went to Melbourne with £500 which he had saved up, and lived with his uncle, James Pearce. "For 3 months or so, I was chiefly interested in seeing and hearing all that was best in the theatre, the art galleries and in music. By this time I had decided to train myself for a career as a Reciter."

Alexander kept himself in Melbourne by taking various jobs—he mentions an estate agency, a department store and a firm of tea merchants—whilst he studied the violin and was "much occupied in the evenings in acting and producing plays."

His interest in reciting increased—"during this time, I did all that was possible to prepare myself for a career as a Reciter, with a small selected Repertoire of dramatic and humorous pieces". But he was already much worried by "hoarseness and lowered vitality", and he had begun his research into his USE: "I considered that the source of my trouble lay in the USE of my vocal organs."

In spite of his vocal trouble, he embarked on a career as a reciter and appears to have had a considerable immediate success, performing in various cities in Tasmania and Australia. But he was increasingly hampered by his voice, and began to search for a method to prevent his trouble. A full account of his research appears in his book *The Use of the Self* and in *The Alexander Journal*, No. 7, 1972.

Once he had cleared up his vocal trouble, Alexander settled in Auckland, where he combined his reciting career with teaching the method he had discovered—"employing a technique that enabled one to change and improve USE". During his time in Auckland he decided to make the teaching of his method his main career and he moved to Melbourne in 1894 to do this.

He continued to give recitals in Melbourne and built up a teaching practice, giving both individual and group instruction. In 1899 he moved to Sydney, where he became firmly established as a teacher of his method—"my teaching was basically one of changing and controlling reaction"—and he became director of the Sydney Dramatic and Operatic Conservatorium between 1900 and 1904.

One of his friends in Sydney, Dr McKay, encouraged him to go to London to teach his method. The decision to move was not easy—"I would be taking a serious risk in an attempt to build up a practice in London, where competition was considerable and where I would be unknown." In 1904—with no quick air or sea travel—Australia was a long way indeed from England both culturally and in actual miles. Nevertheless, "the desire to take the plunge grew rapidly and I decided to take it."

It was a large gamble to take, and he was immediately in some financial trouble—trouble which was with him to some extent for most of his life. Perhaps unfortunately he became interested in betting—horses had always been a passion with him—and on one occasion when "the financial aspect was the one remaining difficulty", he was given a tip for a double on the Newmarket Handicap and the Australia Cup at 150–1, which won him £750 when he was down to his last £5. This appears to have been a turning point in getting himself settled in London.

In the years before World War I, he worked mainly with actors—Henry Irving, Viola Tree, Lily Langry, Constance Collier, Oscar Asche, and Matheson Lang were among his pupils—and he wrote his first book *Man's Supreme Inheritance*, which was published in 1910.

From 1914 to 1924 he spent half of his time in the US and half of his time in England, writing two further books *Conscious Control* and *Constructive, Conscious Control of the Individual*, which had a preface by John Dewey, the American educational philosopher.

In spite of his considerable teaching success, his money troubles appear to have continued—perhaps connected with his habit at this time of spending the months between May and October "resting and racing"! When I first met him in 1938, he was an undischarged bankrupt, although maintaining his teaching practice at 16 Ashley Place in London. I accepted at the time that he had become a bankrupt rather than pay a debt which he considered unfair, but he never quite sorted out the economics of his role in a satisfactory manner, and he tended to get himself involved in a complicated network of financial and social transactions.

After World War I, he was fortunate in having the help of two teachers, Ethel Webb and Irene Tasker. A class for children was started in 1924 in London, and subsequently a school in Penhill in Kent, where his method was made fundamental to the whole school curriculum.

In 1930 a teachers' training class was established in his rooms in London at 16 Ashley Place, and it was carried on there until 1940 when he took his school to America. Whilst in America, his fourth book, *The Universal Constant in Living*, was published.

His time in the US from 1940 to 1943 was not a very happy one. It looked at that time as though his work and his Principle could easily become lost—nearly all of his teachers were in the services.

Eventually Alexander could stand the separation from England no longer, and he returned to London in the summer of 1943. He was now becoming an old man and he was embittered by the refusal of medicine and education to recognise his ideas. The very qualities which had led him to his scientific discoveries—single-mindedness and questioning—tended now towards suspiciousness. This was not helped by an attack on his work in South Africa, culminating in a libel action which he brought successfully against the South African Government in 1948.

He won large damages in the case, but the summing up was fair: briefly that his method was sound but that his presentation of it was misleading. Fortunately the judges saw past his mode of presentation to the value of what he was actually doing.

Alexander in his old age retained to the last his immense teaching skill and patience: but in the years before his death, his hard life had taken its toll, and he began to despair of his ideas ever being accepted without being watered down.

After he returned from the US to England in 1943, a teachers' training course was started once more in London in 1945. In 1948, as noted by Sir Stafford Cripps at a dinner which was later held to celebrate Alexander's eightieth birthday, "there was set up the Society of Teachers, as a central body for the furtherance of the teaching and for the upholding of proper professional standards. To these teachers, Alexander has entrusted the carrying-on of his work in the form in which he has passed it on to them."

Alexander died in 1955, still at work, at Ashley Place. The subsequent years have seen an increasing interest in his Principle. Unfortunately, some people who claim to be teaching it will not necessarily have undergone an adequate training. Membership of the Society of Teachers of the Alexander Technique (STAT) or the possession of a certificate signed by Alexander is proof that a three-year training has been taken at a training establishment approved by the Society. If there is any doubt about the qualification of a teacher, it would be prudent to enquire about this.

These brief biographical details can be amplified by reference to back numbers of *The Alexander Journal* (published by STAT) which contains articles by many of the teachers and doctors who knew Alexander well. Perhaps the clearest account ever given appears in the Memorial lecture which my wife, Marjory Alexander Barlow, gave at the Medical Society of London in 1965 and this was subsequently reprinted by the Alexander Institute.

No one person can give a full account of Alexander. Most people made contact with him through the medium of his teaching, i.e., mainly through non-verbal communication. So much depended upon his presence, his hands, and the whole circumstance of his re-educational situation that one might well despair of giving a clear account of him to a third person. The memory of the man resides in the skill which he was able to pass on to succeeding generations of teachers.

GLOSSARY

BEHAVIOUR

1. *Stimulus/Response:* Lifting a telephone receiver after hearing its bell ring.
2. *Operant:* Lifting a telephone receiver to call someone.
3. *Directed:* Letting telephone ring a little longer, whilst pausing to check over-quick reflex reaction, and then attending to the USE whilst lifting the receiver and responding.
4. *Feed-back:* Allowing behaviour to be influenced by knowledge of results rather than by expectations.
5. *Conditioned:* Behaviour adopted in the past by conscious choice or training but which has become unconscious and unchosen.

BODY-CONSTRUCT

A person's perceptive framework (made up of previous learned experience) by means of which he *construes* what he perceives and edits it to obtain (by his muscular USE) his preferred perception.

CORE-STRUCTURE

A directed USE in which the co-ordination of the head, neck and back is not disturbed by the limbs, the breathing, the use of the special senses, the act of communicating or the various biological functions.

DIAGNOSIS

1. *Descriptive:* The classification of reports on events in a person's past and present life.
2. *Prescriptive:* A diagnosis which carries implications for preventive action.

DIRECTING

Used in two senses:

1. Projecting verbal directions which have both an anatomical and a functional component, e.g., "Shoulder Release". The anatomical direction ("Shoulder") is a signal to put attention on the shoulder: the functional component ("Release") is a signal to release any faulty USE which the anatomical direction may have made apparent.
2. Pointing in a certain direction, e.g., the shoulder should release in a sideways "widening" direction, and not by being dropped inwards across the back of the chest.

DOING

Making unnecessary muscle-tension in an attempt to "do" the Alexander directions. Non-doing = allowing the muscles to act in response to direction. Undoing = releasing unnecessary muscle tension.

DYSTONIA
: USE which involves excessive and/or wrongly distributed muscular tension.

ELECTROMYOGRAPHY
: The recording of small amounts of muscular activity by placing electrodes in or over a muscle.

END-GAINING
: Carrying out actions to attain ends without attending to the USE (the "means") and without gauging the amount of pressure and tension required to start a given activity. On a larger scale, any form of behaviour which does not permit the feed-back of information except that which relates to the one specific end desired.

ERGONOMICS
: The study of USE and posture in a working or social environment, and the design of equipment to facilitate functioning.

HOMEOSTASIS
: The capacity of certain bodily systems to achieve a relatively steady neutral balance, with a greater or lesser degree of oscillation around a central resting point, to which the system seeks to return after stress.

INHIBITING
1. Checking over-quick reflex activity, to give the possibility of choice.
2. In a more sophisticated Alexander sense, it means deliberately taking a moment to stop and say "No" to an action which one has decided in advance to carry out.

KINAESTHESIA
: The perception of USE.

KYPHOSIS
: An excessive "hump" anywhere between the lower neck and the middle back.

LORDOSIS
: An unduly arched-in lower back.

MIS-USE
: USE which involves an inappropriate relationing of one (or many) parts of the body to the rest of the body, either through too much or too little muscle-tension.

NEGATIVE FEED-BACK
: The elimination of error by means of a servo-mechanism.

POSTURE
: The external shape of the body, at rest and in movement, as observed by another person. Produced by the effect of good or bad USE on the genetic physique.

PRIMARY CONTROL
Alexander's phrase to describe the relationing of the skull to the neck, and of the skull-and-neck to the rest of the body.

PRINCIPLE
A hypothesis which is adopted as being logically prior to a range of other questions. An answer which is not itself contained in some prior answer.

REFLEX
A bodily reaction which is almost invariably initiated by the same type of stimulus.

SCOLIOSIS
A sideways curvature of the spine.

SERVO-MECHANISM
A system which is influenced by the consequences of its own behaviour.

SPONDYLOSIS
Arthritis of a part of the vertebral column.

STIMULUS
An event which is followed by a behavioural response.

USE
The characteristic and habitual way of using and moving the body. The relationing of one part of the body to another part in response to circumstances and the environment.

REFERENCES

1. W. Barlow, "An Investigation into Kinaesthesia", *Medical Press Circular* 215, 60, 1946.
2. C. Sherrington, *Man on his Nature*, Cambridge University Press 1951.
3. G. Ryle, *Collected Papers*, Hutchinson 1971.
4. W. Cannon, *The Wisdom of the Body*, Kegan Paul 1932.
5. D. Morris, *The Naked Ape*, Jonathan Cape 1967.
6. Sir A. Keith, "Man's Posture: Its Evolution and Disorders", *British Medical Journal* I, 451, 1923.
7. E. Hooton, "Why Men behave like Apes and Vice Versa", *Science* 83, 271, 1936.
8. W. Le Gros Clark, *The Antecedents of Man*, Edinburgh University Press 1959.
9. B. Campbell, *Human Evolution*, Chicago University Press 1967.
10. A. Schopenhauer, *The World as Will and Idea*, Kegan Paul 1883.
11. H. Spencer, "Gracefulness" in *Essays*, Williams 1857.
12. W. Barlow, "Postural Homeostasis", *Annals of Physical Medicine* 1, No. 3, 1952.
13. H. J. Eysenck, *Dimensions of the Personality*, Routledge and Kegan Paul 1947.
14. R. A. Granit, *The Basis of Motor Control*, London and New York Academic Press 1970.
15. T. Roberts, *Basic Ideas in Neurophysiology*, Butterworth 1966.
16. P. A. Merton, "Nervous Gradation of Muscular Contraction", *British Medical Bulletin* 12, 214–18, 1956.
17. D. E. Broadbent, "Introduction", *British Medical Bulletin on Cognitive Psychology* 1971.
18. W. Barlow, "Posture and the Resting State", *Annals of Physical Medicine* vol. II, No. 4 1954.
19. R. M. Hare, *Freedom and Reason*, Oxford University Press 1963.
20. W. R. Gowers, "Lumbago: its lessons and analogues", *British Medical Journal* 1, p. 117, 1904.
21. J. L. Halliday, "Psychological factors in Rheumatism", *British Medical Journal* 1, p. 213, 1937.
22. I. P. Ellman, "Fibrositis", *Annals of the Rheumatic Diseases* 3, p. 56, 1942.
23. P. Hench, "The Management of Chronic Arthritis and other

Rheumatic diseases among soldiers of the U.S. Army", *Annals of the Rheumatic Diseases* 5, p. 106, 1946.
24. W. Barlow, "Anxiety and Muscle-Tension Pain", *British Journal of Clinical Practice* vol. 13, No. 5, p. 339, 1959.
25. H. J. Eysenck, "Dimensions of the Personality", *op. cit.*
26. W. Barlow, in *Modern Trends in Psychosomatic Medicine*, Butterworths, 1954.
27. W. Barlow, "Postural Deformity", *Proceedings of Royal Society of Medicine* vol. 49, No. 9, p. 670, 1956.
28. W. Barlow, "Anxiety and Muscle Tension", *British Journal of Physical Medicine* 10, 81, 1947.
29. A. Gregg, "What is Psychiatry?", *British Medical Journal* 5, i, 551, 1944.
30. H. Wolff, *Headache*, Oxford University Press 1948.
31. Davis, Buchwold and Frankmann, "Autonomic & Muscular Responses and their relation to simple Stimuli", *Psychological Monographs* No. 405, p. 69, 1955.
32. J. E. Goldthwaite, *Body Mechanics*, Lippincott 1952.
33. C. Darwin, *The Expression of the Emotions in Man and Animals*, Reprinted: Thinkers Library, Watts 1945.
34. M. Bull, "Attitude Theory of the Emotions", *New York Nervous Disease Monograph* 1951.
35. W. Young, *Eros denied*, Corgi 1967.
36. L. Corbin, *Avicenna*, Routledge and Kegan Paul 1960.
37. H. Gardner, "Figure and Ground in Aesthetic Perception", *British Journal of Aesthetics* 1972.
38. Prof. T. Sparshott, *An Enquiry into Goodness*, University of Toronto Press 1958.
39. J. M. Tanner, *Growth at Adolescence*, Oxford, Blackwell Scientific Publications 1962.
40. W. Barlow, "Rest and Pain", *Proceedings of IV International Congress of Physical Medicine, Excerpta Medica International*, Series 107, p. 494, 1964.
41. M. Swain, "Survey of Physical Education", *Australian Journal Physical Education*, June 1961.
42. J. Dewey, *Experience and Nature*, Open Court 1925.
43. Aldous Huxley, *Ends and Means*, Chatto & Windus 1937.
44. F. M. Alexander, *The Use of the Self*, Methuen 1932.
45. Prof. G. E. Coghill, *Anatomy and the Problem of Behaviour*, Cambridge University Press 1929.
46. Judson Herrick, *G. E. Coghill: Naturalist and Philosopher*, Chicago University Press 1949.
47. W. Barlow, "Psychosomatic Problems in Postural re-education", *The Lancet*, p. 659, 2 Sept., 1955.
48. Marghanita Laski, *Ecstasy*, Cresset Press 1961.

49. Mair & Bannister, *The Evaluation of Personal Constructs*, Academic Press 1968.
50. Aldous Huxley, "End-gaining and Means Whereby", *Alexander Journal*, No. 4, p. 19, 1965.
51. B. Pasternak, *Dr Zhivago*, Collins 1958.

INDEX

INDEX

by H. E. Crowe